FROM THE GREAT WALL
TO THE GREAT COLLIDER

From the
Great Wall
to the
Great Collider

China and the Quest to Uncover the Inner Workings of the Universe

S<small>TEVE</small> N<small>ADIS</small>

S<small>HING</small>-T<small>UNG</small> Y<small>AU</small>

 International Press
www.intlpress.com

From the Great Wall to the Great Collider:
China and the Quest to Uncover the Inner Workings of the Universe

by Steve Nadis & Shing-Tung Yau

Copyright © 2015 by International Press of Boston, Inc.
Somerville, Massachusetts, U.S.A.
www.intlpress.com

ISBN: 978-1-57146-310-4

First edition 2015

Printed in the United States of America.
Printed on acid-free paper.

Library of Congress Cataloging-in-Publication Data is available.

19 18 17 16 15   1 2 3 4 5 6 7 8 9

To physicists striving to uncover the laws of nature on scales ranging from the very large to the very small.

And to people throughout the world who support those efforts, pushing in their own ways for the general advancement of knowledge.

CONTENTS

PREFACE

Living on the Edge of the Math–Physics Interface

Many who think of me strictly as a mathematician might be surprised to hear that I am actively encouraging China to build what my colleagues and I sometimes refer to as the "Great Collider"—a term first suggested by David Gross, a Nobel Laureate in Physics. The device in question could be the largest and most powerful particle accelerator that humankind has ever seen, depending on what happens to similar plans being hatched elsewhere. The point of such an undertaking is not to build such a machine simply for the bragging rights of having "the world's biggest," but rather to assemble an instrument capable of opening up new frontiers in fundamental physics, while unlocking secrets of the universe that were previously beyond our reach.

I'm interested in this endeavor for numerous reasons, many of which are laid out at length in this book. The primary motivation of my co-author and I was to see whether a plausible case could be made for this gargantuan and extremely challenging construction project. We, along with many of my colleagues in the field, have determined that a machine of this scale is essential to further our progress in basic physics and, perhaps more importantly, to satisfy humanity's long-standing desire to understand the world around it.

Rather than summarize that case here, I will instead say something about my own motivation. First off, I should point out that although most of my work might fall under the heading of "mathematics," I have published dozens of articles in physics journals, and I hold appointments in both the math and physics departments at Harvard. The reason for this overlap really comes down to a matter of personal preference: I like working at the boundary of these two dynamic fields. For one thing, I find it an exciting place to be. But beyond that, I've come to realize that ideas from physics—inspired by observations of nature and the quest to fathom it—can be very stimulating for mathematics.

It cuts both ways, of course, as mathematics is vital to physics on many levels. I'm not just talking about the fact that the laws of physics are, in essence, mathematical formulations. Advances in mathematics often have strong implications for physics and sometimes show up in nature as concrete physical manifestations. The two fields are intertwined to such a degree that it's often hard (and not especially fruitful) to say where the mathematics ends and physics takes over.

Perhaps an anecdote will make things clearer. When I arrived in the United States in 1970 to pursue graduate studies in mathematics at Berkeley, I became fascinated with Einstein's theory of general relativity, which I hadn't studied before in any depth. My specialty back then was, and is still to this day, geometry, and Einstein offered us something new—a geometric conception of gravity. Rather than picture the phenomenon as an attractive force between massive objects, he said, it's better to think of gravity as the curvature of space-time due to the presence of massive objects. The warping of space-time, which Einstein spoke of, relates to its so-called "Ricci curvature." That got me thinking: What if the space under consideration was a vacuum, wholly devoid of matter or anything else to give it heft? Might there still be curvature (of the Ricci sort) even in the absence of mass?

To my surprise and delight, I soon discovered that almost the very same question—expressed, almost unrecognizably, in rather abstract mathematical language—had been posed two decades earlier by the geometer Eugenio Calabi in a conjecture that bears his name. Calabi maintains, however, that when he unveiled his the conjecture in 1953, "it had nothing to do with physics," at least in his mind. "It was strictly geometry."

The Calabi conjecture called for the existence of even-dimensional geometrical shapes or "spaces" endowed with a peculiar kind of symmetry, among other features. Many geometers considered such entities "too good to be true" and believed that spaces satisfying the conditions set by Calabi were mathematically impossible. Although I was initially a skeptic, I worked many years on the Calabi conjecture and ended up proving in 1976 that it was true. The spaces that were theorized by Calabi, and whose existence I eventually affirmed, came to be called Calabi-Yau spaces or manifolds. I strongly sensed that my work could be important for physics—apart from the questions about general relativity that I started with—but I didn't know exactly where or how it might fit in.

At roughly the same time, starting in the early 1970s, physicists began speculating about a hypothetical symmetry of nature called "supersymmetry," which—if true—could solve many puzzles associated with quantum field theory, the prevailing theory of particle physics (at that time and to this day). No one has yet shown whether our universe is, in fact, supersymmetric, but Calabi-Yau spaces possess this kind of symmetry, which means the idea has at least passed the test of mathematical plausibility. Some of its implications for physics will be explored later in this book.

But the story doesn't end there. In 1984, I was contacted by physicists who were looking for higher-dimensional spaces imbued with supersymmetry to play a central role in string theory, and they thought Calabi-Yau spaces might be the ticket. String theory attempts to do what no theory in physics has successfully pulled off before—explain and unify all the forces and all the particles of nature. We're still a long way from knowing whether string theory is the correct description of our universe, and no experiments are likely to settle this question anytime soon. That said, string theory has already stimulated important mathematical work on multiple fronts. For example, physicists hit upon a previously unknown kinship between different Calabi-Yau spaces that's called "mirror symmetry"—a finding that resuscitated the field of enumerative geometry, leading to the solution of problems that in some cases dated back to the 1800s. My colleagues and I attempted to trace this notion to its roots, thereby helping to establish the mathematical underpinnings of mirror symmetry.

As the foregoing example shows, there's a steady flow running back and forth between physics and math that is beneficial, as well as critical, to both fields. I've found, time and again, that the intuition of physicists can be extremely helpful to mathematicians. I'm also aware of many instances when work in mathematics has aided the cause of physicists. Apart from my own selfish motivations for supporting physics, I happen to really like that field. I'm a big proponent of physics and, indeed, of all science at the frontiers.

But fundamental physics, which supplies a stream of ideas to mathematics, cannot run indefinitely without some periodic input from actual experiments. To put it bluntly, physicists cannot be sure their ideas are right until they are confirmed empirically. I still vividly recall my 1990 visit to CERN, Europe's big high-energy physics center based in Geneva, where I was shown the circular tunnel—27 kilometers in circumference and about 100 meters underground—in which particles would eventually be accelerated in the Large Hadron Collider (LHC). I was truly impressed by the engineering, which provided exquisite precision on a mammoth scale. A couple of decades later, the LHC delivered on its great promise, putting the finishing touches on the so-called "Standard Model" of particle physics with the discovery of the Higgs boson, sometimes referred to—in colloquial terms—as the "God particle." (Scientists, however, almost never call it that, regarding the expression as more of a marketing ploy than anything having to do with science.)

The time has now come for a new machine, bigger and better than its predecessor—this time with a tunnel perhaps 100 kilometers around so that particles can be accelerated even faster still, resulting in collisions at five to ten times greater energies than before. Such a facility could carry us to a whole new level of insights beyond the Standard Model, possibly fostering the discovery of a host of new particles along the way. It's really not overstating things to call the instrument needed to get us there the Great Collider. It is, I admit, a grandiose term, but this is the sort of project that actually warrants such exalted language.

Given that I was born in China and have devoted much energy toward boosting scientific and mathematical research there—including the founding of six mathematical institutes in China, Hong Kong, and Taiwan—I would love to see this colossal particle accelerator erected on my native soil, even though its construction and operation will necessarily be an international endeavor in almost every sense. While the machine may be based in China, it would be intended for everyone to use.

Rather than sitting back and hoping that something happens, I decided to take a more active role. I have met with physicists in China, the United States, and Europe, and with top Chinese officials to discuss the potential benefits of this undertaking. One point that I often stress is that this project has the potential to lift *all* science—in the country and outside the country as well—and not just in those areas related to physics. This could happen in the same way, as the saying goes, that a rising tide can lift all boats. Although one extremely prominent Chinese leader did not give the proposal an outright yes, the response was not negative either. And that's actually quite a triumph at this stage since it keeps the idea alive, and gives proponents like me hope that we might actually pull this off, while buying some additional time in which to do so.

I also moderated a panel discussion on this subject in Beijing with some of the world's leading physicists, which drew appreciable attention to our proposition from all quarters. Afterwards I helped get a letter about the collider into the hands of another official within the highest ranks of the Chinese government. The latter task was not as easy as it sounds because prior to delivering the letter to him, I first had to discuss the issue with the head of the Chinese Communist Party's Organization Department, the Minister of Science and Technology, and the president of the China Association for Science and Technology. Each of these steps was crucial given that any one of these powerful individuals was in a position to deal a major (if not fatal) blow to the project, but they instead agreed that the idea was worth considering, overriding some influential critics who have opposed the collider from the outset.

On a personal level, I'd like to see China enhance its stature on the world's scientific stage, and, frankly, that stature could use some enhancing. Even though government spending in science and technology has risen dramatically in recent years—and the number of papers published by Chinese scholars has grown accordingly—the quality and originality of those papers and the research behind them, as reflected by citations and other metrics, is still lagging compared to the output of the United States and other Western powerhouses. It's fair to say that in particle physics and other disciplines of science, China has some catching up to do—and, in fact, a lot of catching up to do. In some areas, we may actually be decades behind.

Nevertheless, beyond my parochial sentiments on this matter, I'm of the opinion that efforts like this accelerator venture—which have the capacity to inspire people—transcend national boundaries and are important for everyone on this planet. The machine I'm talking about here would not just be China's collider but would rather be a collider for the whole world, involving physicists drawn from every quarter of that world.

This proposition would tie into a broader enterprise that goes to the heart of what the term "civilization" is all about. Simply put, human societies need to keep exploring in all kinds of directions—geographic, scientific, and artistic—in order to stay healthy and stay relevant. I also believe that when a goal is truly worthwhile, history invariably favors those who do what it takes to get things done—not the naysayers who focus on a project's high price tag and other shortcomings, seeking out reasons for *not* doing things rather than trying to move forward.

The kind of experiments discussed in this book, made at this proposed new mecca for high-energy physics research, will undoubtedly break through boundaries that constrain us today, thereby enhancing our grasp of the universe's inner workings. And to me, the knowledge gained in this process would rank among humanity's greatest possible achievements. Frankly, I can't think of anything higher to strive for.

—*Shing-Tung Yau, Cambridge, Mass., 2015*

In August 2014, I went to China for the first time; it was my first trip to Asia as well. I'd previously toured all over the United States—a nation with a "back story" of just a few hundred years. China's culture and history, by contrast, date back thousands of years. It's for that reason that I often associate China with what used to be called the "Old World." Yet during my visit, I found a country—with bullet trains, maglev travel, and cutting-edge research in genetics and genomics—that was in some ways more modern than the much newer nation I had come from. And I also found, somewhat ironically, people in the so-called Old World who were making great strides in charting the new world and future of high-energy physics—a field that directly and indirectly drives practically every other area of science. Ambitious plans of this type are not being laid, in any serious fashion, in my own country, I'm disappointed to report.

In China, I was greeted with many new vistas and stunning scenery, and a variety of unfamiliar foods, as well as great warmth and hospitality among the people I met. In my travels from Beijing to Hong Kong and many points along the way (as well many points *not* along the way), I encountered dozens of individuals who were eager to aid my research. I'd like to thank them for their assistance, plus others in the United States, Europe, and elsewhere in the world, without whose help this book would not have been possible. Here's a list of some of the people who have lent a hand to this endeavor, though I apologize in advance to anyone whose efforts were overlooked:

Jon Butterworth	Chao-Lin Kuo	Nathan Seiberg
Pisin Chen	Ken Lai	Shu-Heng Shao
Weiren Chou	Xiaonan Li	Xiaoyan Shen
Ming-Chung Chu	Ying–Hsuan Lin	Gary Shiu
Robbert Dijkgraaf	Michael Loy	Matt Strassler
John Ellis	Cai-Dian Lu	Andy Strominger
Jie Gao	Wang Lu	Meng Su
Brian Greene	Kam-Biu Luk	Gerard 't Hooft
Lijun Guo	Luciano Maiani	Yanjun Tu
Fred Harris	Masahiro Morii	Edward Witten
Hong-Jian He	Qing Qin	Chenghui Yu
Rebecca He	Christopher Rogan	Chuang Zhang
Ren Juan	Matthew Schwartz	

I'd like to give special thanks to seven individuals who provided a tremendous amount of help: Nima Arkani-Hamed, David Gross, Joe Incandela, Ashutosh Kotwal, Michael Peskin, Henry Tye, and Yifang Wang. They generously devoted their time and expertise to this project for one reason only: because they care about the future of high-energy physics and want to see their field advance and thrive.

Hong-Jian He and Zhong-Zhi Xianyu deserve special commendation for taking on the task of translating our sometimes baroque English prose into flawless text for the Chinese-language edition of this book. Chenfang Wang of IHEP has done a fantastic job in tracking down photos of Chinese physicists and research facilities. Brian Bianchini of International Press has taken our drab-looking Word documents and magically converted them into a polished book, with Aileen McElroy handling publicity efforts and Lixin Qin overseeing the whole endeavor. In addition, we're delighted with the sensational cover art provided by Xixi

Huang. We'd also like to recognize Ying Lin for the work he did indexing the book.

Throughout this process, Maureen Armstrong, Lily Chan, and Irene Minder have offered invaluable administrative support. Gary Malmon was kind enough to show me around Beijing (sometimes at what seemed like grave peril on the back of his motor scooter!). And finally, I'd like to thank my family members—Melissa, Juliet (who helped with the cover design), and Pauline; and my parents, Lorraine and Marty—for their steady encouragement during the year I worked on this book.

My coauthor, S.T. Yau, is also grateful for the support provided by his wife, Yu-Yun, and his sons, Isaac and Michael.

—*Steve Nadis, Cambridge, Mass., 2015*

PROLOGUE

The Next Great Wall

THE GREAT WALL OF CHINA—a ring of fortifications built to protect China from Mongol invaders and other would-be plunderers from the north—is aptly named and well deserving of its imposing title. Visible to Earth-observing satellites and commonly cited as one of the "Seven Wonders of the World," the Great Wall is the longest structure ever fabricated by humans, as well as one of the most ambitious architectural and engineering projects ever carried out. More than 6,000 kilometers of the wall, most of which were built during the Ming Dynasty (1368-1644), presently stand—newly restored in some stretches and severely eroded in others. The wall's total length, including faint archaeological remnants that date back more than 2,000 years, was recently measured at more than 21,000 kilometers. [1]

Composed of rock, bricks, cement, and mud, the singularly named "wall" is more accurately described as an overlapping and sometimes discontinuous network of walls, interspersed with periodic watchtowers and sentry posts. The main wall, plus its numerous branches and spurs, stretches from the East China Sea to the Jiayuguan Pass in the northwestern province of Gansu, following ridgelines and crossing innumerable mountains along the way. Breathtaking scenery draws hordes of sightseers to this celebrated attraction, with about 4.5 million tourists per year visiting the Badaling portion of the wall alone, located some 70 kilometers north of Beijing. [2] Other sections are equally picturesque but more remote, passing through barren, rocky terrain, far from cities and modern-day crowds.

Construction of the Great Wall, according to the historical consensus, was initiated by the first emperor of a unified China, Qin Shi Huang, who lived from around 260 to 210 B.C., although the structure incorporated some walls that were erected as far back as the seventh century B.C. The edifice was originally known as the "Long Wall of Ten Thousand Li,"[3] which technically implies a length of 5,000 kilometers (a *Li* being equal to 500 meters), but figuratively means "endless." The enterprise that he started wasn't quite endless for Emperor Qin, who died at the age of 50, about ten years into the construction, despite his attempts at gaining immortality. About one-fifth of China's population at that time took part in the construction effort,[4] and the wall largely fulfilled its original purpose—as it did in centuries to come—serving as an effective barrier against trespassers from the north.

Now, some 2,200 years later, the geopolitical circumstances are rather different. China no longer needs a ring of fortifications to fend off marauders advancing by land with malevolent designs. Rather than serving any defensive purposes, the formidable barricade erected in centuries past is now a sight to behold, a place for contemplation, and one of the world's leading tourist destinations. What's being considered today is a project of almost comparable magnitude—a Great Wall for a new era, which will incorporate the latest technological advances to propel science towards uncharted frontiers and, hopefully, carry the human spirit with it. The idea, on a conceptual level, is to take the magnificence of that millennia-old edifice and apply it in scientific rather than militaristic directions.

Plans specifically call for building the largest machine ever assembled—a "Great Collider" that will hurl tiny particles nearly at the speed of light, circling through a tunnel about 100 meters underground and up to 100 kilometers in circumference, in order to unlock the secrets of the universe. A device of this sort would not come cheap, costing perhaps one ten-thousandth of China's gross domestic product during the years the facility is being built and operated. But the collider could confer broad benefits to society and serve as a tremendous source of national pride, just like the Great Wall itself. "And hopefully its discoveries—and its contributions to the development of science and technology, and to China in general—will be remembered just as long," says Physics Nobel Laureate David Gross of the University of California, Santa Barbara, who has come to China on several occasions to talk about the project with physicists and government leaders.[5]

MIT physicist Frank Wilczek, who collaborated with Gross on their Nobel Prize-winning work, concurs with this view. Although he was not referring specifically to China's plans, Wilczek called a giant particle accelerator that was previously in the works a towering "monument to curiosity."[6]

And that, indeed, is the overriding purpose of the newly proposed collider—to satisfy humanity's never-ending quest for knowledge, shedding light on questions that have likely been pondered, in some fashion, since the beginning of conscious thought. Simply put, the primary objective of this undertaking is to provide insights into the workings of nature at its most basic level—at the level of the particles that make up every object in the world around us and the forces that guide every interaction taking place therein.

The technology required for an endeavor of this order will likely spur long-term development in China, sparking residential and commercial development in areas surrounding the new laboratories. Some 10,000 or so scientists and engineers—and an additional number of construction workers—will eagerly come to this region from all over the globe to participate in an effort that lies on the forefront of science. In this respect, the new collider will serve almost the opposite function to that which the Great Wall has had—at least in terms of that structure's original purpose. "The Great Wall was meant to keep foreigners out, while the Great Collider is meant to bring foreigners in," notes the physicist and cosmologist Henry Tye, director of the Institute for Advanced Study at the Hong Kong University of Science and Technology.[7] Encouraging outsiders to come into China can, perhaps ironically, make the country more secure (again contrary to the Great Wall's initial premise) through the international partnerships as well as friendships forged, and the spirit of connectedness thereby instilled.

The big draw for outside scholars relates, in some ways, to the depth and reach of physics itself. It is really no exaggeration to characterize physics as the most fundamental of sciences—the "science of everything," as it's sometimes called—because all the other fields—be they biology, chemistry, geology, oceanography, or meteorology—are ruled, at their roots, by the laws of physics. And if plans for this new science complex are realized, the ensuing activity would place China at the world's epicenter of high-energy physics research for many decades. The facility, in fact, will help sustain, and may even save, the field of high-energy physics that has contributed so much to our understanding of the cosmos.

Coincidentally (or perhaps not so), the prime site under consideration for the mammoth accelerator lies near Qinghuangdao, a port city and popular beach resort about 300 kilometers east of Beijing, which is often referred to as the "starting point of the Great Wall." It's the place where the Ming Dynasty-era wall, which is shaped in the form of a dragon's head, juts out into the Pacific. That's as far as this vast complex extends. And from there, this pier-like promontory reaches out and—as an inscription reads—"shakes hands with the sea."

The construction and successful operation of the proposed physics center near the Great Wall's eastern border could, as David Gross suggested, signal the start of something equally momentous, if not more so, inaugurating a bold intellectual inquiry that could help people grasp not only their place in the firmament but the very makeup of the firmament itself. Achieving that will take a machine of unprecedented capabilities—a Great Collider indeed.

A section of the Great Wall of China, north of Beijing, overlooking the Huanghuacheng reservoir (Photograph courtesy of Ken Lai)

INTRODUCTION

A Clarion Call

AS A PHYSICIST WITH A THRIVING RESEARCH CAREER and as director of the Institute of High Energy Physics (IHEP) in Beijing—a center within the Chinese Academy of Sciences that has more than 1,400 employees and an annual budget in excess of $200 million (US)—Yifang Wang has a lot on his mind. But of all the issues Wang has to grapple with, one is paramount: Does high-energy physics, the subject around which this entire research operation revolves, actually have a future? Both as a scientist and as an administrator, Wang is working hard to ensure that it does.

With regard to China alone, the country's main particle accelerator—the Beijing Electron Positron Collider II, which is situated on the IHEP campus—should approach the end of its useful research life by the early-to-mid-2020s. Therefore Wang, the winner of the 2014 W.K.H. Panofsky Prize in Experimental Physics and other awards, wants to make sure that, at a minimum, Chinese physicists have a high-quality machine on which to ply their trade. But his concern extends well beyond his nation's borders. The world's premier machine of this sort, the Large Hadron Collider (LHC) in Geneva, Switzerland, won't be around forever either. Once its work wraps up in a couple of decades, the field as a whole could shut down unless, somewhere, a bigger and better replacement comes to the fore.

So far, nothing is firmly in the works. And it will take decades of planning, construction, instrument tuning, and debugging to get a future accelerator, capable of carrying us to the frontiers of physics, ready for operation. In other words, someone needs to start making preparations for the next high-energy physics machine, which could be a couple of decades in coming. Wang believes that that "someone" could be China, and he has spent the past several years mobilizing physicists in his country and elsewhere, as well as conferring with political, civic, and business leaders, in the hopes of turning this wild notion into reality.

It is not pure fantasy, given that China has already made an indelible mark on modern physics. Chen Ning (C.N.) Yang and Tsung-Dao (T.D.) Lee won the 1957 Nobel Prize in Physics for their groundbreaking insights into an asymmetry, or "parity violation," associated with the weak interaction—a basic force of nature that is involved in radioactivity, nuclear fusion, and other processes. Samuel Ting, who is also of Chinese descent, shared the 1976 Nobel Physics Prize for his discovery of a new kind of elementary particle, the J/Psi. These pioneers (whose work is described in greater detail in Chapters 1 and 4) helped lay the foundation upon which the current generation of Chinese scholars now depend, reminiscent of Isaac Newton's famous quote about "standing upon the shoulders of giants."[1] While the contributions of Yang, Lee, and Ting have been hugely important for the development of particle physics, as well as for the creation of a homegrown physics establishment in China, much of their seminal work has been done overseas. Many Chinese physicists, Wang included, are hoping the day has come when researchers don't have to go abroad to do cutting edge work—a day when some of the main action in their field is taking place right inside their own country.

Halfway around the globe, Nima Arkani-Hamed at the Institute for Advanced Study in Princeton was harboring similar thoughts. Arkani-Hamed, a 2012 Fundamental Physics Prize winner who is considered one of the leading theorists of his generation—"the one we all look up to," according to fellow theorist David Kaplan of Johns Hopkins[2]—was becoming increasingly dismayed about the state of physics in the world and particularly in the United States, which has been his home for the past 20-plus years. "We as a field still suffer from SSC Post-Traumatic Stress Disorder," he says, referring to the Texas-based Superconducting Super Collider, which was slated to be the world's largest and most powerful particle accelerator until the U.S. Congress halted construction and cancelled the project in 1993.

That, coincidentally, happened to be the same year in which Arkani-Hamed entered graduate school at Berkeley; and ever since then, he says, "American particle physicists keep settling for less and less."[3] Our horizons are shrinking at a time they should be expanding, he adds. Congress is perhaps even more dysfunctional and shortsighted now than it was in 1993, lacking virtually any appreciation of the value of basic physics research. "We need to find a way to change the game somehow," Arkani-Hamed urges, "or our generation could be the one that dropped the ball in the development of the subject that's been handed down to us from the likes of Kepler, Galileo, Newton, Einstein, and Dirac. We could be the ones that [screwed] it up, and that's what keeps me awake at night."[4]

With the United States' profile receding and the European economy flat at best (and their hands full, in any case, running the LHC), Arkani-Hamed had wondered who would take the initiative and build the next big accelerator. But a couple of years ago, an idea popped into his head and wouldn't go away: "*China should build it.*" He'd been talking up this notion—"What about China? What about China?"—whenever he could and to whomever would listen to him, but those discussions remained at the late-night "bull session level" until the summer of 2013 when Arkani-Hamed heard that almost identical rumblings were emanating from within China itself.

He went there in August 2013 to see for himself, spending a long time at IHEP talking with Yifang Wang and his colleagues. Arkani-Hamed was impressed by what he saw and heard. "There are people here who recognize how hard it will be," he said at the time, "but are willing, nevertheless, to roll up their sleeves and try to make it happen." He had come to China to see whether or not it was "a total pipe dream" and became convinced that the effort had a real chance for success—a good enough shot for him to sign on as head of the brand new Center for Future High Energy Physics. The Center, which is based within IHEP, was officially formed in December 2013 but had informally begun during conversations among Wang, Arkani-Hamed, and other physicists in August of that year. From that point onward, a growing core of scientists has embraced a common goal: to start the process that will hopefully culminate in the construction of a giant new particle accelerator, which would surpass the LHC and ill-fated SSC—the former being a site of magnificent triumph and the latter a testament to failure, a partly-dug tunnel to nowhere, buried deep within the Texas soil.

For the most part, these researchers hoped that the next big machine would be built in China, drawing on the best expertise available, both within the country and without. But they also recognized that the most important thing was that a machine of this sort be built somewhere, anywhere in the world, to keep the field alive and sustain the longstanding quest for knowledge at the most fundamental level.

To drum up support for this dream, leaders from the world of particle physics met in Beijing in February 23, 2014 at Tsinghua University for a panel discussion led by one of the authors of this book, Shing-Tung Yau. The group that gathered at Tsinghua University included Wang and Arkani-Hamed, as well as such luminaries as Joseph Incandela, David Gross, Luciano Maiani, Hitoshi Murayama, Gerard 't Hooft, and Edward Witten—winners of the Nobel Prize, Fundamental Physics Prize, Dirac Medal, Fields Medal, and other prestigious awards. The scientists had come to Beijing to discuss the future of their field and to make the case for the next-generation machine that could take us there—China's proposed "Great Collider."

The university auditorium in which this idea was introduced to the Chinese public was "a madhouse," Arkani-Hamed reports. "There were hundreds and hundreds of students, people hanging from the rafters, and amazing levels of enthusiasm. There is this enormous population of talented people here, and it would be amazing to tap into it."[5]

Physics Nobel Laureate David Gross told the audience: "There is a wonderful opportunity for China to step up and take a leading role in the exploration of the most basic questions in the physical sciences. This is a great adventure that some of us have come to help support."

The enterprise is really quite essential, he stressed, offering the only means yet devised that can keep this field moving forward. "As usual, we suspect that nature is smarter than we are and that we'll have to get direct clues to figure out how it works," he says. "The easiest way to do that, of course, is to look, to observe, and to experiment, and that's what we intend to do."[6]

The strategy makes perfect sense to Edward Witten, a renowned theorist who laid some of the critical groundwork for string theory and M-theory, among other notable contributions. "In the long run, theoretical physics slows down without new inspiration coming from experiments," Witten said. "In the last few decades, we've been blessed with surprises and great discoveries like the accelerated expansion of the universe. For the coming period, we also need to pursue the experimental frontier in order to keep supplying the inspiration that the theorists need."[7]

Luciano Maiani, the former director-general of CERN—the celebrated laboratory that runs the LHC—told the students packing the hall that he hoped that IHEP's "well-set strategy to explore the future will be realized. And I hope that China will join, in this way, the U.S., which has made the Tevatron [particle accelerator at Fermilab], and Europe, which has made the LHC, in the club of nations that lead the search for beauty and simplicity at small distances." CERN has been a tremendous success story, Maiani added, "a story of money invested that has yielded great returns to the government in terms of image, technology, and the training of young people."[8] China, he suggested, could be well-situated to reap some of these same benefits too—as well as other benefits yet to be seen.

Hitoshi Murayama—a University of California-Berkeley physicist and director of Japan's Institute for the Physics and Mathematics of the Universe—picked up on the latter theme. Scanning the crowd of animated and engaged students, Murayama said that probably the most important thing about hosting a big project in a given country is that so many people can potentially be excited. "This is very important for the future of the country. If young people don't get excited about basic science, mathematics, and engineering, then there is no future for the country. And that kind of excitement can really change the country and also the world."[9]

There are many sensible reasons for pursuing a project like this, along with many problems in physics that need to be addressed. But the most compelling reason for getting involved in such an endeavor, according to Arkani-Hamed, is the spirit of adventure."[10] He realizes, of course, that it won't be all fun and games. "As Kennedy said about going to the moon: We choose to do these things not because they are easy but because they are hard." That's the sense we seem to have lost, and the spirit he hopes we'll soon regain.

The conversation among the distinguished panelists at Tsinghua was a triumph in terms of fueling interest in the new collider and the far-reaching exploration it would support. Prior to this event, these ideas had mainly been of interest to a relatively small group of particle physicists. But thanks to the intense media coverage, the conversation had been extended to include the general public and politicians as well.

A few months later, the symposium participants and several other Nobel Laureates in Physics wrote a joint letter, which Yau—the letter's first signatory—delivered to a top Chinese official, after first running the proposal by several key party leaders. In their letter, Yau and his fellow authors described the idea behind the new accelerator, arguing that "China is perfectly positioned to host this machine and spearhead the international effort." Moreover, they said, "Thousands of the world's most talented physicists and engineers would flock to China to enthusiastically join in the effort." As a result of building the collider, the scientists added, "China will catapult into global leadership of fundamental physics in the 21st century."

The project's backers had gotten the shot in the arm they had hoped for. They also knew that this preliminary stage—imagining *what could be*—was the easy part, buoyed by a wave of exhilaration they had helped to unleash. The really hard work, in terms of calculating, planning, deliberating, and eventually carrying out this massive undertaking, was yet to begin. And the Center for Future High Energy Physics, where much of these preparations were to be hashed out, was bound to be a busy place for years to come. The next step was to write a detailed report on the experimental goals for this machine, working out in advance the range of measurements that the new collider would make possible.

There were many other items on the physicists' agenda: They needed to conduct geologic studies to determine the best site for the circular tunnel (up to 100 kilometers in circumference), develop more powerful magnets for bending and steering particle beams, build detectors of unprecedented accuracy, and create enhanced methods of information processing, storage, and transmittal. Identifying critical technical issues that require research and development, prior to proceeding to the design phase, was also a big priority.

Another crucial task would be to establish the chief questions in physics that the collider will address—and then to determine just what the answers to those questions might tell us about the universe, which we hadn't previously known. No matter how much theorists are able to work out in advance, they know that a new machine that delivers a big jump in energy is also bound to deliver a slew of surprises that may lead to insights surpassing anything they ever anticipated. And that's what they're hoping for—to be surprised, and maybe even shocked, because that's what ultimately contributes to the greatest leaps in understanding.

But before venturing too far into the discussion of the future of physics, and what this new collider might someday uncover, it may be instructive to say something about this field's illustrious past. Such an exercise will hopefully give readers a sense of how we got to where we are today, so that they can have a better grasp of where we might go next in this grand excursion to the edge of the known and beyond.

In February 2014, an illustrious group of scientists met in Beijing to make the case for a giant particle collider that would open up new frontiers in physics. Participants in this symposium included (as seen from left to right) Shing-Tung Yau, Hitoshi Murayama, Joseph Incandela, Edward Witten, David Gross, Gerard 't Hooft, Nima Arkani-Hamed, Luciano Maiani, and Yifang Wang. (Photograph courtesy of IHEP)

Chapter 1

Smashing Atoms

THE QUEST TO UNDERSTAND THE BASIC CONSTITUENTS OF NATURE—an effort that China's proposed collider would help further and indeed "accelerate"—has its origins at least as far back as the 5th century BC. For it was in that distant era, in the coastal village of Abdera in northeastern Greece, that the philosopher Leucippus and his student Democritus developed their theory of "atomism." The premise of atomism, an idea that also took root in India around the same time, holds that everything we can see and touch—all matter, in other words—is composed of small, indivisible elements called atoms. These atoms are the building blocks out of which more complex structures such as planets are built, in the same way that lavish palaces can be built out of lowly bricks, elaborate computer programs out of 0's and 1's, and timeless literary masterworks out of mere letters.

The general notion advanced by Leucippus and Democritus—that matter is made up of small, standardized bits (like the individual pieces in a Lego set)—turned out to be correct, although it remained in the realm of philosophy and metaphysics for more than two thousand years because the ancient Greeks did not have the technical wherewithal or the analytical methodology needed to verify this proposition or even to attempt doing so. Chemists like John Dalton made empirical advances in the early 1800s, which bolstered the atomist interpretation. But it was not until later in that century and in the early 1900s that physicists started acquiring the technology that would unequivocally confirm that hypothesis—while also showing that atoms were not, as originally supposed, the smallest unit of matter below which no further cuttings or subdivisions were possible.

A breakthrough came in 1897 in experiments conducted by the British physicist J.J. Thomson at Cambridge University's Cavendish Laboratory. Thomson was studying the mysterious "rays" produced by a new kind of technology, cathode ray tubes—elongated glass vacuum cavities that create a discharge when a voltage is applied to electrodes at opposite ends. Thomson concluded that the rays given off by these devices were, in fact, a stream of particles that he called "corpuscles." He measured the charge-to-mass ratio of these corpuscles— or electrons, as they came to be known—and identified them as components of atoms that could be stripped free, thus challenging the long-held conviction that atoms were unalterable and inviolable. In this way, Thomson inaugurated the search for "subatomic" particles—some fraction of which are classified as "elementary" or "fundamental" after scientists conclude that they appear to have no internal structure and are not made up of even smaller, simpler units. Electrons, based on everything we've learned so far, fall into the latter (elementary) category.

New, compelling evidence for atoms themselves came a few years later. In 1905, Albert Einstein published a paper, which explained that Brownian motion—the apparently random motions of dust grains floating in water, first observed by microscope almost 80 years earlier by Robert Brown—was caused by individual water molecules bumping into the grains. Einstein's analysis and mathematical arguments gave strong credence to the notion that matter was indeed composed of atoms and molecules, as had long been supposed. In experiments that charted the movement of particles suspended in liquid, the French physicist Jean Perrin showed that Einstein's theoretical calculations were right on the mark, thereby demonstrating—to the satisfaction of the scientific establishment—the existence of atoms and molecules. Perrin also made reasonably accurate estimates regarding the size of these entities, receiving a 1926 Nobel Prize for his efforts.

Now that it had been determined that the atoms the ancient Greeks had speculated about were in fact real, Ernest Rutherford, a British physicist born in New Zealand, resolved to learn about their nature and underlying structure, assuming there was any structure to be found. The prevailing view at the time, the so-called "plum pudding" model proposed by Thomson, who supervised Rutherford's post-graduate research, held that atoms were mushy, permeable balls— spheres of positively-charged matter into which negatively-charged electrons were interspersed like plums mixed into a pudding. In experiments that started in 1909 at the University of Manchester, Rutherford and his colleagues decided to put this idea to a test.

The researchers used a radioactive source—a sample of the gas, radon-222— to emit a stream of alpha particles, which are helium atoms stripped of their two electrons. The particles were directed towards a sheet of thin gold foil. A screen set up behind the foil would emit a flash of light when struck by an alpha particle. If atoms—gold atoms, in this case—were permeable like pudding, the alpha particles would pass through the gold foil without resistance and hit the screen. But some of the particles were instead deflected at large angles, occasionally bouncing straight back. "It was almost as incredible as if you fired a fifteen-inch shell at a piece of tissue paper and it came back to hit you," Rutherford commented.[1] He concluded that the atom's positive charge and most of its mass were tightly concentrated in the center—in what we now call the nucleus—rather than being evenly distributed, as had been previously supposed. This compact nucleus, in the newly emerging picture, was surrounded by a much more diffuse swarm of electrons. And comparatively speaking, the nucleus is rather minute—10,000 to 100,000 times smaller than the atom it sits in—while comprising more than 99.9 percent of the atom's mass.[2]

Owing to findings like this, which quite literally got to the heart of matter, Rutherford firmly believed that physics—in contrast to fields such as botany and zoology—was the most fundamental of subjects. "All science is either physics or stamp collecting," he asserted in a statement that probably earned him little love among non-physicists.[3]

But Rutherford was not done making fundamental contributions to his field. He continued to turn out new scientific insights while promoting the advent of technology that would help probe the structure of the atomic nucleus, which he had just discovered. In 1919, before moving to Cambridge University to head the Cavendish Laboratory, Rutherford bombarded nitrogen nuclei with alpha particles, dislodging positively charged particles that he identified as hydrogen nuclei. The same thing happened in subsequent experiments when he hit other elements, such as oxygen and aluminum, with alpha particles. Hydrogen nuclei again shot out, and Rutherford surmised that they must be basic constituents of all nuclei and hence fundamental particles in their own right. He called them protons.

In 1920, Rutherford predicted that the nucleus must also contain particles of neutral charge, or neutrons, because the nucleus would be unstable if it consisted only of positively charged protons. He encouraged researchers at the Cavendish to look for neutrons. At the same time, he started promoting the advent of man-made devices that could produce particles in greater number, and of greater energy, than those obtained from natural radioactive sources. A machine of this sort, he told the British Royal Society in 1927, could be useful in a broad range of physics experiments, including the neutron search he was then championing. "It has long been my ambition to have available for study a copious supply of atoms and electrons, which have an individual energy far transcending that of the alpha and beta particles from radioactive bodies," Rutherford said in an address to the Society over which he presided from 1925 to 1930. "I am hopeful that I may yet have my wishes fulfilled, but it is obvious that many experimental difficulties will have to be surmounted before this can be realized on a laboratory scale." [4]

As it turned out, he did not have to wait long for two of his dreams to come true, at least in part. In 1932, James Chadwick, a physicist in the Cavendish Lab, announced the discovery of the neutron. It had thus become clear that the atom, far from being indivisible, consisted of "nucleons" (protons and neutrons), electrons, and—as remained to be seen—perhaps even smaller ingredients that might someday be laid bare.

In that same year, John Cockcroft and Ernest Walton, who also worked under Rutherford in the lab, succeeded in splitting a lithium nucleus to create two alpha particles. Of perhaps greater importance was how Cockcroft and Walton achieved this: They used an electrical potential difference of 800 kilovolts to accelerate protons through an eight-foot-long vacuum tube, placing a lithium target at the end of the tube. They not only were the first to "split the atom," but they had also built—and employed—one of the world's first particle accelerators. [5] Devices of this sort were the "microscopes" that physicists would continue to employ in their attempt to uncover the structure of matter at ever finer scales.

Across the ocean at roughly the same time, Ernest Lawrence at the University of California, Berkeley, was fashioning a new kind of particle accelerator that would ultimately have a big impact on the field. Inspired by a paper written by the Norwegian engineer, Rolf Wideröe, Lawrence invented a novel, circular accelerator, which he called a "proton merry-go-round" but later became known as the cyclotron. The first cyclotron—built in 1931 using a combination of glass, sealing wax, and bronze, with some household items thrown into the mix—was only about five inches in diameter. It used a magnetic field to force charged particles onto a circular path; while an electric field sped the particles up to higher velocities as they repeatedly passed through the same accelerating region. The device succeeded in boosting the energy of protons to 80,000 electron volts or eV. (An electron volt is the amount of energy gained, or lost, while moving an electron across an electric potential of one volt.) An eleven-inch diameter cyclotron built later that year, mainly by Lawrence's assistants, increased proton energy to more than one-million electron volts. Five years later, a cyclotron built in Lawrence's lab accelerated alpha particles to energies of 16 million electron volts. [6]

The era of accelerator physics was thus born. Accelerators have become bigger and more powerful ever since, with many of the world's largest devices being direct descendants of Lawrence's cyclotrons. The physicist Edward Witten explains the push to bigger machines: "Many of nature's secrets are revealed only when subatomic particles collide at extremely high energies," he says. [7]

Through this approach, researchers have been able to reproduce, ever so briefly, the conditions they believe prevailed shortly after the Big Bang—a time when the elementary particles were thought to have been created. Conditions would have been so hot in this early epoch, in fact, that only the most fundamental particles would have existed. Which is one reason why particle physics is such a keen area of interest. It offers a way—surely one of the best paths identified to date—for learning about our universe's first moments.

Going to higher energies also enables physicists to probe nature at smaller distance scales. This fact is a consequence of a hallowed tenet of quantum physics, the Heisenberg uncertainty principle, which holds that if you want to examine something tiny like a particle, the uncertainty of your measurements in terms of distance will be inversely proportional to the particle's momentum. In other words, the greater the momentum, the greater the precision of your distance measurements. It therefore takes high energies and high momenta to be able to accurately study nature on small physical scales. That's just what accelerators are good for, and it's essentially what they're built for.

The hope, notes Nobel Prize-winning physicist Steven Weinberg, is that by moving to the new levels of energy that particle accelerators open up, new phenomena will be illuminated. "This expectation has almost always been fulfilled."[8]

The 20th century was, in keeping with Weinberg's statement, an incredibly productive era in physics during which 200 or so particles were observed, measured, and labeled. The field responsible for most, if not all, of these discoveries is called "high-energy physics or particle physics," says Witten, but physicists aren't solely preoccupied with particles themselves. "What we're really trying to do is to understand the most fundamental laws of nature."[9] Physicists, in other words, want to identify the smallest building blocks out of which the tremendous variety of structures—from tiny to vast in scale—arise. They also want to learn about the forces that hold these pieces together, push them apart, or affect them in other ways.

Particle accelerators have been vital to this process ever since their introduction in the early 1900s. They have been used with great success to confirm theoretical predictions regarding the existence of various particles. And they have also delivered great surprises, which can be the biggest payoff of all—finding something that no one expected to see or would have thought to look for in the first place.

Most modern accelerators still borrow from Lawrence's original design. Like the earlier models, they take a charged particle and use magnetic fields to keep it moving in the proper direction (i.e., to "steer the beam") and electric fields to boost a particle's speed as high as possible (or desired). They then slam it into a fixed target or collide it with another particle. (The latter approach, involving head-on collisions, offers a practical way of reaching higher energies and is therefore generally preferred, although it requires much better aim.) Detectors surrounding the point of impact sift through the collision debris and record the outcome of the event.

Accelerators generally come in two forms (although facilities often use some combination of the two): Particles can be confined to a circular course, gaining speed each time around the "racetrack"—the so-called synchrotron approach, which is a descendant of Lawrence's earlier cyclotrons. Or particles can be sent down a linear track in which they are shot out at one end of a tunnel and travel to the other end, gaining speed as they go along. In either case, the objective is to smash particles together, or into a fixed target, and see what emerges from that collision—an experimental approach that is simple in concept yet usually employs the most sophisticated technology devised up to that time.

The point of such an exercise is not merely to see what a particle is made of—like throwing a clock against a wall to try to figure out its various parts. The goal is typically more ambitious than that: It's to make something new, something that has never been seen before. And when two particles, each carrying a lot of kinetic energy, crash into each other, they can combine to create a more massive particle (that is almost always unstable). The original particles' masses plus their kinetic energy can be converted into the new particle's mass in accordance with Einstein's famous formula: $E=mc^2$. Total energy is conserved in this transaction, which is required by the laws of physics, but mass is not. This is allowed because Einstein's formula shows that mass is just another form of energy, meaning that the energy of motion can be transformed into the energy of mass and vice versa. It doesn't matter which form the energy takes so long as it all adds up to the same original sum.

Investigators are eager to move to higher and higher energy collisions because it gives them a chance to study particles that have never been observed before—the higher the collision energy the heavier the particles that might spring into existence.—though we have to be quick to catch them, as these newly created particles are not normal habitués of our everyday world. They tend to be short-lived, typically sticking around far less than a microsecond, and they move very fast. "The experiments are like entering a lottery," explains Oxford physicist Frank Close. "Most of the collisions produce familiar particles, but once in a while you get lucky and something unexpected shows up." [10]

Of course, new particles don't announce their presence for all to see. Instead, physicists have to comb through the debris left over from a collision, sifting for clues like detectives scouring a crime scene. They may never actually see the new particle itself but instead observe the *effects* of that particle—the debris left in its wake. Researchers employ a range of detectors to facilitate their task—devices that can record, for instance, the tracks that particles leave and determine their speed, mass, and charge, doing the same for the particles' offspring, or "decay products," as well. After extracting as much information as possible from a single event—and more often from a very large number of events—analysts, aided by high-powered computers, can try to get a fix on the identity of the various players in the drama they've witnessed and hopefully stumble upon a novel suspect or two. In many cases, new particles are not "seen" directly, but their existence can still be inferred, "beyond reasonable doubt," from the weight of multiple strands of evidence stacked up on top of each other.

In this way, generations of researchers have followed up on the work of Thomson, who discovered the electron, and of Rutherford, who discovered the proton and oversaw the discovery of the neutron. One question, which became important in the early 1930s, was whether the proton and neutron had their own "antimatter" counterparts, the antiproton and antineutron. The antiproton, for example, would have the same mass as an ordinary proton but would have opposite (negative) electric charge and magnetic moment (or magnetic polarity). Because a neutron has no net electrical charge, the antineutron would have the same mass and charge but an opposite magnetic moment.

Speculation about antimatter was sparked by a theory presented in 1928 by the British physicist Paul Dirac that combined Einstein's special relativity with quantum physics in order to describe the behavior of fast-moving ("relativistic") particles. The Dirac equation predicted the existence of a particle with the same mass as an electron and a charge of the same magnitude but of opposite sign. His theory stated, in other words, that there should be positively charged electrons or "antielectrons"—a notion that many physicists considered implausible.

But in 1932, Carl Anderson of the California Institute of Technology identified that very thing, a positively charged electron, among the showers of particles produced when cosmic radiation hits the Earth's atmosphere. The cosmic rays entered a cloud chamber that Anderson had built—a container filled with a super-saturated vapor that was subjected to an intense magnetic field. Charged particles follow a curved path in the presence of a magnetic field and the degree of curvature depends on a particle's mass. The trajectory of the new particle spotted by Anderson had the same curvature as an electron, but it curved in the opposite direction implying that it had an equal mass as well as an equal and opposite charge. He called it the positron, and the name stuck. This was the first bit of antimatter ever detected, although Anderson was initially unaware of Dirac's earlier prediction on this subject. [11] Anderson won a Nobel Prize in Physics for this work in 1936—the same year he discovered another particle, the muon, whose existence had not been predicted in advance and whose role in the grand scheme of things was not readily apparent. (This unexpected finding prompted the Nobel Prize-winning physicist, Isidor Rabi of Columbia University, to famously quip: "Who ordered that?" [12]). The muon had a charge of minus one, exactly the same as the electron although the new particle was 207 times more massive.

Dirac's theory, however, predicted more than just positrons. Dirac showed that when you combine quantum mechanics with relativity, a new symmetry arises, which implies that practically every kind of particle has an antiparticle of opposite charge (although some particles may be their own antiparticle). His proposition almost doubled the number of particle types to be found in the universe—a notion that seemed farfetched at the time but was eventually shown to be correct.

Following the positron discovery, interest among physicists then turned to antiprotons, which would presumably be a basic building block of antimatter. An antiproton and an antielectron (or positron) could combine, for instance, to create an antihydrogen atom—something that was, in fact, achieved, although not until 1995. But antiprotons, unlike positrons (and muons), could not be found through cosmic ray studies and must instead be produced in the laboratory.

The catch was that in the 1930s no particle accelerator was close to being energetic enough to produce antiprotons, which would require the simultaneous creation of a proton-antiproton pair. A proton and an antiproton each have the same "rest mass"—which is to say the minimum value of energy these particles possess when completely at rest—of about one billion electron volts or, specifically, 938 million electron volts (MeV). So the accelerator would have to produce a minimum of twice that energy or about 2 billion electron volts. In that era, the unit of a billion electron volts was called a BeV although it's now referred to as a GeV (or gigaelectron volt).

Ernest Lawrence, at a facility later named the Lawrence Berkeley Lab in his honor, initiated the construction of a machine that would soon be called the Bevatron—a circular, synchronized proton accelerator or synchrotron capable of producing 6.5 billion electron volts in collision energy. When it was switched on in 1954, the Bevatron stood as the biggest and most powerful particle accelerator that had ever been built. The machine was constructed largely for one purpose—to make antiprotons—and physicists wasted no time in setting out to do just that.

Within a year, a team headed by Berkeley physicists Emilio Segrè and Owen Chamberlain had found a total of 60 antiprotons, negatively-charged particles with the same mass as a proton, unambiguously identified through measurements of their momentum and velocity. For each antiproton that the researchers had spotted, 40,000 other particles created in this process had to be weeded out. [13] The team announced its discovery on October 19, 1955. Just one year later, Bevatron scientists led by Bruce Cork, produced the first antineutrons ever seen by passing a beam of antiprotons through ordinary matter.

In that same year, 1956, there was an even more momentous event in the world of theoretical and experimental physics: Two Chinese-born physicists, Tsung-Dao Lee and Chen Ning Yang, published a paper in *Physical Review*, which argued on theoretical grounds that the laws of nature are not always symmetrical. [14] The issue at stake here was the conservation of parity—a revered principle of 20th century science holding that a physical process, and the reflected image of that same process as viewed through a mirror, should operate by the same

Portrait of J. J. Thomson
(Photograph courtesy of the Chemical Heritage
Foundation, Philadelphia, Pennsylvania)

Ernest Lawrence, founder of the Berkeley Lab, won
the 1939 Nobel Prize for Physics for "the invention
and development of the cyclotron."
(Photograph courtesy of the Berkeley Lab)

Ernest Rutherford at work in his McGill University laboratory, *circa* 1907
(Photo credit: W.H. Hayles / McGill University Archives, PR0017771)

The Berkeley Lab's 184-inch cyclotron, *circa* 1942 (Photograph courtesy of the Berkeley Lab)

"Drift tubes," which are key elements of the linear accelerator of the Bevatron-2545, *circa* 1961 (Photograph courtesy of the Berkeley Lab)

rules and yield the same outcome. Put in other terms, physical reactions should show no preference for left over right or for right over left. It was believed, in fact, that nature could not distinguish between the two.

Lee and Yang—who were then based, respectively, at Columbia University and Princeton's Institute for Advanced Study—challenged that venerable assumption. They concluded that parity violations could be ruled out in strong and electromagnetic interactions because such violations would have been readily apparent from previous data. But the data concerning the weak interaction was still inconclusive, Lee and Yang said. And they proposed several weak interaction processes that might be checked for parity conservation, so that this matter—long accepted as gospel—could finally be subjected to an experimental test.

One of the examples they suggested related to two subatomic particles that had recently been produced in accelerators, the tau and theta. These particles looked identical in almost every respect: They had the same mass and spin, and their lifetimes—that is, how long they stayed intact before decaying into other particles—were essentially the same as well. But the tau decayed into three smaller particles called pions, whereas the theta gave rise to two pions. Owing to these different end-of-life outcomes, physicists had assumed that the tau and theta must be different particles. Lee and Yang offered an alternative interpretation: If the weak interaction did not conserve parity then the tau and theta could be the same particle that simply decayed in different ways. Parity violation, if confirmed, could thus resolve the tau-theta puzzle. And if parity could be violated in this case, Lee and Yang said, it could be violated in other instances as well.

Empirical corroboration came quickly, and Lee and Yang became the first scientists of Chinese descent to win the Nobel Prize for discovering the first documented case of an asymmetry in nature. The initial confirmation was the indirect result of a conversation between Lee and a physics colleague at Columbia who also came from China, Chien-Shiung Wu, regarding possible experiments that might bear on the matter of parity conservation. Wu, who has been called "the Chinese Madame Curie" and "the First Lady of Physics Research," devised an experiment that she performed in the summer of 1956 with colleagues at the National Bureau of Standards (NBS) in Washington—a facility chosen because it had the capacity to chill materials to extremely low temperatures. [15] Wu and her fellow investigators cooled the radioactive isotope, cobalt-60, to one-hundredth of a degree above the absolute zero. The material was placed in a strong magnetic field, which caused all the cobalt-60 nuclei to line up and spin in the same direction. The researchers then carefully observed the electrons given off during the isotope's nuclear decay, which was also governed by the weak force. If parity were conserved, the same number of electrons would fly off in the direction of the nuclear spin as those flying off the opposite way. Wu detected a clear breach in parity, as more electrons were emitted in a direction opposite to that of the spin. "When Dr. Wu knocked out that principle of parity," wrote the author Clare Boothe Luce, "she established the principle of parity between men and women." [16]

While Wu and the NBS physicists were verifying their results, a separate example of parity violation was confirmed almost simultaneously at Columbia's Nevis Cyclotron by Richard Garwin, Leon Lederman, and Marcel Weinrich, who studied the decay of a muon into an electron and two neutrinos. Their paper appeared, back to back, with the paper by the Wu group in the same 1957 issue of *Physical Review*. Muon decay experiments at the University of Chicago's Synchrocyclotron, carried out by Jerome Friedman and Valentine Telegdi, also showed parity violation. Friedman and Telegdi's paper came out in the next issue of *Physical Review* (also in 1957), possibly because it arrived at the journal one day later, having been sent from Chicago to the New York-based publication. [17]

Taken together, these experiments demonstrated, beyond dispute, that reactions governed by the weak force and their mirror images do not have to abide by the same rules. This discovery brought the notion of symmetry breaking into fundamental physics—a development that has since proven to be hugely consequential. But initially this finding caused some bewilderment. "A rather complete theoretical structure has been shattered at the base, and we are not sure how the pieces will be put back together," commented Isidor Rabi. [18]

In the meantime, more examples of parity violation kept popping up. The phenomenon was observed later in 1956, for instance, at the Bevatron when researchers collided protons with negative pions and monitored the decay of the resultant (A hyperon) particles. Lee and Yang had proposed this very experiment, and the Berkeley team succeeded, becoming the first to demonstrate parity nonconservation in the decay of hyperons—particles classified as "baryons," which are discussed later.

Before long, Berkeley physicists participated in another discovery that ultimately led, through a complex chain of events, to something very big. During an experiment at the Bevatron, they created a fleeting combination of three particles—consisting of an L hyperon and two pions—which lasted a mere 10^{-28} seconds. This ephemeral "particle," the first so-called "strange resonance" ever observed, was denoted $Y_1^*(1385)$—the number referring to its rest energy of 1385 million electron volts. Although other resonances had been seen before, including "the famous Fermi 3,3 resonance," explained the physicist Walter Alvarez, who led the research, "the impact of the Y_1^* resonance on the thinking of particle physicists was quite different; the Y_1^* really acted like a new particle." Alvarez and his colleagues announced the discovery of the Y_1^* at the 1960 Rochester High Energy Physics Conference, he recounted, "and the hunt for more short-lived particles of this sort began in earnest." [19]

Tsung-Dao Lee and Chen Ning Yang working together at the Institute for Advanced Study. (Photo by Alan Richards, courtesy of the Shelby White and Leon Levy Archives Center, Institute for Advanced Study, Princeton, New Jersey)

Ernest Ambler of the National Bureau of Standards and Chien-Shiung Wu of Columbia University, following their successful experiment in 1956, which demonstrated the nonconservation of parity as predicted by T.D. Lee and C.N. Yang (Photograph courtesy of NIST archives)

By the end of 1960, members of the Berkeley group had discovered two more resonances, K*(890) and Y*(1405). From the masses of these and other resonances, physicists were able to estimate the mass of another hypothetical particle, dubbed the omega-minus, whose existence had been independently predicted in the theories of Caltech physicist Murray Gell-Mann and the Israeli physicist Yuval Ne'eman. However, producing the omega-minus, with an expected mass of 1676 million electron volts, would require more energy than the Bevatron could furnish.

In 1960, the Alternating Gradient Synchrotron (AGS) at Brookhaven National Laboratory in New York supplanted the Bevatron as the world's premier accelerator, delivering a new record energy of 33 billion electron volts (GeV). The omega-minus particle was discovered at the AGS in 1964. Its existence, along with the existence of other, previously seen resonances like Y_1*(1385), lent strong support to the quark model proposed in 1964, again independently, by Gell-Mann and George Zweig who was then a visiting scientist at CERN (the particle physics research complex centered around Geneva, Switzerland).

Gell-Mann and Zweig's theory pertained to hadrons—a broad class of particles that included baryons, such as protons and neutrons; and mesons, such as pions. All hadrons, they argued, are composed of smaller, more fundamental constituents called quarks, a term that Gell-Mann took from James Joyce's novel, *Finnegan's Wake*. Quarks are unlike other known particles in that they have a fractional electron charge equal to one third or two thirds that of an electron and proton.

Hadrons have two other main constituents in addition to quarks: antiquarks, the antimatter counterpart of quarks, and gluons, particles associated with quarks that will be discussed later in this chapter. The models advanced by Gell-Mann and Zweig initially called for three different quarks, each with its own antiquark, although physicists have since identified six quarks in total, along with six corresponding antiquarks. All baryons (protons and neutrons included) are made up of three quarks, and each meson is made up of a quark and antiquark.

When Gell-Mann and Zweig put forth this idea in the early 1960s, particle physicists were trying to make sense of all the hadrons that were being discovered in large numbers, with no apparent end in sight and no obvious scheme for classifying them. The quark model helped bring some order to this seeming chaos because it explained that all the hadrons that had been identified up to that time—and all those that have been seen ever since—were different combinations of the same three quarks and antiquarks.

The model had two features that struck many observers as curious. One was the notion of fractional charges, which had never been seen before in nature. The second puzzling aspect stems from the fact that no individual quarks have ever been seen. There are, moreover, theoretical reasons for believing that quarks can never be seen roaming freely, on their own, even though this was a novel concept in particle physics.

"The reason a quark cannot be isolated is similar to the reason that a piece of string cannot have just one end," explains Brookhaven physicist Michael Creutz (although he concedes that "one can't have a piece of string with three ends either"). [20] Instead, the arguments go, quarks can only exist in combinations, permanently "confined" within hadrons. Assuming that supposition is true, as certainly appears to be the case, the dedicated experimentalist still has ways of catching a glimpse.

In fact, landmark experiments were carried out in the late 1960s and early 1970s to probe the inner structure of protons—just as Rutherford had probed the inner structure of the atom 60 years earlier to show it had a nucleus, and had then demonstrated with colleagues some decades hence that the nucleus, itself, had smaller components called protons and neutrons. But this next round of experiments, aimed at investigating the proton's internal makeup, required a newer and bigger machine. In 1962, groundbreaking began at the Stanford Linear Accelerator Center (SLAC). The facility was originally called "Project M," with "M" standing for monster, because it was—and still is—the longest linear accelerator ever built, more than 3.2 kilometers long. The building housing the accelerator is, in fact, the longest in the United States and the third longest structure in the world after the aforementioned Great Wall and the Ranikot Fort in Pakistan—a ringed fortification 26 kilometers in circumference.

In previous experiments that made use of a smaller machine, Stanford physicists had had good luck bouncing electrons off protons in order to measure the proton's size. With a far more powerful electron beam at their disposal courtesy of SLAC—a beam that would eventually reach 50 GeV—the physicists hoped to go even further and perhaps get a glimmer of what was lurking inside the proton.

Construction on the new facility was completed in 1967, and experiments led by Richard Taylor of Stanford and Jerome Friedman and Henry Kendall of MIT began that year, running off and on through 1973. The researchers shot beams of accelerated electrons into targets containing liquid hydrogen—the nuclei of which consisted of protons. Just as Rutherford and his coworkers had carefully observed what happened to alpha particles when they collided with gold atoms (and sometimes scattered at large angles), Friedman, Kendall, and Taylor studied the scattering of electrons by protons. At first, it was hard to discern a pattern. "The data were strewn all over the place," Kendall commented. "It looked like chicken tracks all over the graph." [21]

But after Stanford theorist James Bjorken helped the investigators reformulate the data in a different way, the picture became much clearer: Most of the electrons passed through the target pretty much undisturbed, they observed, but significant numbers of electrons were deflected at large angles. The evidence showed that the proton's mass and charge was concentrated into three discrete, point-like lumps, which presumably corresponded to the three quarks that Gell-Mann and Zweig had postulated. It was subsequently determined that the proton consists of two "up" quarks and one "down" quark; the omega-minus particle discovered in 1964 consists of three "strange" quarks. (While the foregoing is true, we now know that protons and other hadrons are somewhat more complicated than this simple picture would suggest—a point that will be touched upon in the next chapter.)

Within a year, similar findings were obtained at CERN in experiments involving the proton scattering of light particles called neutrinos (which will be briefly discussed later in this chapter and somewhat more extensively in Chapter 4). The electron and neutrino results complemented each other well, according to Kendall, and taken together made a strong case for the presence of quarks inside protons. [22]

But widespread acceptance of the quark interpretation still took a few years, owing to the fractional charge issue and the fact that quarks, themselves, had never been seen. The SLAC experiments suggested that stuff was moving around inside the nucleus, but no one knew why that stuff—including quarks—could move around without ever getting outside.

Theoretical insight emerged later in 1973 when David Gross, Frank Wilczek, and David Politzer came up with an explanation for quark confinement called "asymptotic freedom"—work that earned them a Nobel Prize in 2004. At the core of their idea was the notion that the nucleus and its contents are held together by the strong force. This force keeps a firm grip on quarks, acting like a stiff rubber band that becomes stronger as quarks move farther apart. In fact, the force becomes so strong at large distances that quarks—and the force-carrying particles exchanged between them called gluons—can never be pulled far enough apart to become isolated. Instead, the theory maintains, the quarks and gluons are confined to tight quarters inside the proton (and inside hadrons in general). They can move about freely within this cramped region—and at very high speeds. In fact, gluons, being massless particles, move at the speed of light. Nevertheless, they never get out of the nucleus, nor can quarks, because the strong attractive force acting between them keeps any particles from getting away.

Additional help for the quark model came from an unexpected quarter—from the discovery of a new kind of particle, which took most physicists by surprise, based on research that proceeded concurrently on the east and west coasts of the United States. MIT professor Samuel Ting, who grew up in China for almost all of his first 20 years, was skeptical of the gospel that the universe had only three kinds of quarks—the up, down, and strange quarks that had already been discovered by then. He conceived of an experiment to look for more.

The idea wasn't completely farfetched, as the physicists Sheldon Glashow of Harvard and James Bjorken had proposed in 1964 that there should be a fourth quark, which Glashow named "charm," but there was little evidence at that time for the existence of such a quark. In a 1970 paper, however, Glashow, John Iliopoulos, and Luciano Maiani offered an explanation as to why there *had* to be a fourth quark, suggesting a mechanism called "GIM" after their respective last names. The prediction of the charm quark is generally credited to these three physicists.

Ting set out to find the charm quark a couple of years later, though he wasn't getting much encouragement for the search he had in mind, perhaps because many physicists were satisfied that three quarks seemed adequate to explain the phenomena that had been observed so far. And they assumed that a hunt for additional quarks would come up empty-handed.

Ting, a driven experimentalist, was not inclined to hold back on an idea he liked just because some theorists told him not to bother. "I am happy to eat Chinese dinner with theorists," he is reported to have said. "But to spend your life doing what they tell you is a waste of time." [23]

Ting's proposed experiment was turned down by several major labs, but he persisted and finally managed to secure time at Brookhaven's AGS, [24] which had by then been surpassed by higher-energy machines at CERN and Stanford. He began his experiments at Brookhaven in the spring of 1972 and continued for more than two years. In tests he ran there in 1973 and 1974, Ting and his collaborators shot proton beams into a beryllium target, producing secondary streams of hadrons. Ting surmised that a newly created hadron would give off an electron-positron pair during its decay. But in August 1974, he came across an electron-positron pair that stood out from the pack—a real outlier. By carefully measuring the energy of this unusual pair, Ting determined that the mystery particle responsible for its production was heavier than most of the particles discovered up to that time, having a mass of 3.1 GeV, or roughly three times that of a proton. More curious still was the particle's longevity. It "lived" just a tiny fraction of a second, only about seven zeptoseconds (or seven billionths of a trillionth of a second), but that was still thousands of times longer than expected—a finding that was wholly at odds with physicists' understanding of elementary particles. [25] "On Earth, most people live less than 100 years," Ting said. "But suppose you find a village where everybody lives 10,000 years or 100,000 years. These people must have some different properties," and the same could be said about his new particle. [26]

Ting concluded that the unusual properties of the particle he snared at the AGS owed to the fact that it was composed of a never-before-seen quark (later identified as the charm quark) and its antiquark. He named the particle "J" because the first character of his name, as written in Chinese, resembles a "J". [27]

Ting could have been the sole discoverer of the J particle and its associated charm quark but, being meticulous, he insisted on redoing the experiment several times, and in several different ways, to be absolutely certain of his result.

While visiting California in the fall of 1974, Ting was surprised to hear that a group at SLAC, headed by Burton Richter, had discovered a heavy new particle, also with a mass of 3.1 GeV, which they were calling "Psi." Ting and Richter got together on November 11, 1974 in Stanford, and it quickly became apparent that they had discovered the same particle, soon to be renamed J/Psi, along with capturing the first experimental evidence for a fourth quark, the "charm" quark, thus confirming the theory advanced by Glashow, Iliopoulos, and Maiani four years earlier. Ting and Richter shared the Nobel Prize in 1976, with Ting being the first person ever to deliver his Nobel banquet speech in Mandarin. [28]

In some sense that was just the beginning, which was why the November 1974 meeting between Ting and Richter—and the announcement of their joint discovery later that day—is said to have ushered in the "November revolution." That breakthrough was about more than just a single particle. New baryons and mesons were soon predicted, possessing different amounts of charm. Some of these new particles were found at SLAC. A J/Psi relative, the "psi-prime," for instance, was discovered just ten days later, and the only way to make sense of its behavior was through the charm quark. That's why the discovery of the J/Psi particle went far toward cementing the case not just for the charm quark but for quarks in general, showing in convincing fashion that there was yet another layer of structure in matter. And the "revolution" that this finding sparked helped bring order and logic to what had been a baffling menagerie of particles. New classification schemes, along with an overarching theory to explain them, began to take shape.

Now that there was solid evidence for four quarks, physicists were encouraged to seek out more. In 1972, two Japanese physicists (and eventual Nobel Prize winners), Makoto Kobayashi and Toshide Maskawa, made a compelling theoretical case for the existence of two additional quarks, subsequently dubbed the bottom and top. Kobayashi and Maskawa showed that a new quark "family" was needed to explain a broken symmetry between particles and antiparticles that had been first observed in 1964.

Hints that there might be two more quarks waiting to be found also emerged from another line of reasoning, which related to a different symmetry. The discovery of the charm quark in 1974 led many physicists to believe in a symmetry between quarks and leptons, which seemed to come in pairs. Although this link between the different kinds of particles was not well understood at the time, it suggested that the number of quarks and leptons had to be the same. According to this way of looking at things, up and down quarks were related to the electron and the neutrino. The more massive strange and charm quarks, similarly, were related to the muon and the muon neutrino. Confusion arose in 1975 when SLAC physicist Martin Perl discovered a new lepton, the tau, which was 3,500 times heavier than the electron, its apparent cousin. If the previous pattern held, it meant that the tau was part of a third category, or family, of quarks and leptons—sitting at a higher energy scale—whose other members had not yet been identified. Perl was confident of this interpretation, which was why he named the particle "tau (after the first letter of Greek word for "third"), because he believed—in the face of considerable skepticism—that the particle he had identified was, in fact, "the third charged lepton." In 1977, confirmation that the tau really was a lepton, and kin to the electron and muon, started to come in from the DORIS electron-positron accelerator in Hamburg, Germany. [29]

In that same year, researchers at the Fermi National Accelerator Laboratory outside of Chicago (named after the Italian-born physicist Enrico Fermi) discovered a new particle, the upsilon meson, which consisted of a bottom quark and its antiquark, thereby verifying that the first of the two predicted quarks really did exist. The second hypothetical particle, the top quark, was also discovered at Fermilab, but not until 1995.

Most physicists are satisfied that there aren't any additional quarks to be found beyond the six already identified. But back in the 1970s, one salient question still remained: What held quarks together? Clues came from the Friedman, Kendall, and Taylor experiments at SLAC, which showed that, in addition to harboring three, point-like quarks, the proton also contained electrically-neutral constituents, which were later determined to be gluons—or particles so named because they hold (or "glue") quarks together inside protons, neutrons, and other hadrons.

In 1976, three theorists—John Ellis, Mary Gaillard, and Graham Ross—published a paper that laid out a method for finding gluons through the annihilation (or mutual self-destruction) of electrons and positrons. Such collisions, they argued, would give rise to a distinctive signature, so-called "three-jet events," consisting of a gluon, quark, and antiquark flying off in three separate directions from the impact zone. At that time, Ellis said, "no theorist seriously doubted the existence of the gluon, but direct proof of its existence, a 'smoking gluon,' remained elusive"—for at least another couple of years. [30] Meanwhile, another important paper on three-jet events, by Thomas Degrand, Jack Ng, and Henry Tye, came out in mid-1977, followed by one on a similar topic later that year by George Sterman and Steven Weinberg.

Gluons were detected in 1979 in four separate experiments carried out at the newly commissioned PETRA machine at DESY, a German national research center. At that time, PETRA stood as the world's largest and most powerful electron-positron accelerator, capable of generating 40 GeV collisions. [31] The initial results were achieved in June 1979 by members of the so-called TASSO collaboration, which made use of a detector at PETRA of the same name. TASSO researcher Sau Lan Wu and her co-worker Georg Zobernig observed the first "three-jet event"—an affirmation of the general strategy outlined by Ellis and his colleagues, and carried further by other theorists.

The European Physical Society credited the gluon discovery to four leaders of the TASSO team—Wu, Paul Söding, Bjørn Wiik, and Günter Wolf—while also citing the vital contributions of three other groups working at PETRA that summer, including one headed by Samuel Ting. (Incidentally, Wu, who was born and raised in Hong Kong, had previously been part of the Ting-led team that discovered the J/Psi particle.) All four groups saw the three-jet events that proved the existence of the gluon, and all four groups presented their findings at an August 1979 symposium at Fermilab, where the gluon sightings were publicly announced for the first time. [32]

Collectively, the PETRA experiments upheld the notion that gluons are the carriers of the strong force. These particles provide the "glue," in other words, that binds the atomic nucleus, overcoming the electromagnetic force that would otherwise push protons apart, while keeping individual quarks from flying free.

The discovery of the gluon helped build confidence in quantum chromodynamics (QCD) as the accepted theory of the strong interaction. QCD describes the interactions between quarks and gluons and composites of quarks like protons, neutrons, and more unstable particles called mesons. QCD has been described by MIT physicist Frank Wilczek "as an expanded version of QED," or quantum electrodynamics, the theory of the electromagnetic force that holds atoms together and describes how light and matter interact. [33] QED involves interactions of electrically charged particles, and the electromagnetic force between them is transmitted by photons, which are sometimes referred to as particles of light. Electric charge is conserved in all electromagnetic interactions.

In QCD, similarly, the strong force is transmitted by gluons, which are analogous to photons. There is also a form of charge in QCD called "color," which has nothing to do with everyday colors but is directly analogous to the electric charge in QED. Color charge comes in three varieties in QCD and is always conserved in physical processes. The theory of the strong force concerns itself strictly with the behavior of particles, like quarks and gluons, which carry color charge.

QCD, in turn, rests on a broader idea called Yang-Mills theory, which was introduced in 1954 by Chen Ning Yang and Robert Mills, whom Yang had met at Brookhaven in the previous year. A central feature of Yang-Mills theory is the concept of gauge invariance, which means, in extremely simplified terms, that the equations of motions—and, indeed, all important physical quantities—are left unaffected by a change in coordinate systems. An object, for example, will hit the ground at the same speed regardless of whether it's dropped from the roof of a 100-meter-tall building onto the street, or from a 100-meter butte onto a high-elevation mountain plateau. This is connected to the broader notion, also part of gauge invariance, that the laws of physics aren't affected by changes in position (or "translations"), nor are they altered by rotations around any point or axis in space.

In the late 1940s, Yang started becoming intrigued with gauge invariance, which is a guiding principle of both QED and general relativity, the prevailing theory of gravity. He wondered whether it might provide the underpinning for theories of the strong and weak forces, as well. But it was not until 1953, when he teamed up with Mills, that he pursued this idea in earnest. [34]

Their idea of incorporating gauge symmetry led to "a beautiful theory," according to Steven Weinberg, "because the symmetry dictated the form of the interactions." [35] And ultimately the approach paid off, for we now know that the strong force can indeed be described by Yang-Mills theory.

The same is true of the weak force, which also operates inside the nucleus. Unlike the strong force, the weak force does not hold the nucleus together but instead allows nuclei to undergo various changes. The weak force specifically governs the radioactive decay of particles from one form into another, such as the decay of a neutron into a proton, an electron, and a neutrino, or the reverse transformation of a proton into a neutron. Radioactive decay makes plate tectonics possible on Earth, and probably life as well, by keeping the planet's interior hot. The weak force is also responsible for nuclear fusion, the process that makes the sun shine, and for the synthesis of chemical elements inside stars.

But Yang-Mills theory ran into a major snag in February 1954, when the duo was just getting started. The problem came to light during a seminar Yang led on the subject at Princeton. Wolfgang Pauli, one of the principal architects of quantum physics, was in attendance, and he relentlessly grilled Yang about the particles called bosons that would purportedly transmit the strong and weak forces. Pauli had seized upon a sore spot that Yang was already cognizant of: The equations he and Mills had laid down predicted that the strong and weak forces are carried by massless, electrically charged bosons. In fact, gauge symmetry required the bosons to be massless. But particles of this sort have never been observed in nature, and it was assumed that if massless gauge bosons did exist they surely would have shown up before in accelerator experiments. [36] Pauli knew this because he had derived equations similar to those of Yang and Mills. But Pauli did not publish his study, as he explained, since one would always obtain bosons "with rest mass zero." [37] And what good was a theory that attempted to describe nature, if it instead described things that seemingly could not exist?

Yet the mathematics was so beguiling that Yang and Mills could not and would not abandon their theory. They still hoped to salvage it by finding a way around the problem of massless bosons associated with the strong and weak in-

Aerial view of the two-mile-long SLAC linear accelerator in Menlo Park, California
(Photograph courtesy of the SLAC National Accelerator Laboratory)

Henry Kendall, Jerome Friedman, and Richard Taylor gathered in Stockholm to receive the 1990 Nobel Prize in Physics (Photograph by Lars Åström, courtesy of the Nobel Foundation)

Robert Mills, *circa* 1999, in his office at Ohio State University (Photograph courtesy of the Ohio State University Archives)

Samuel Ting at Brookhaven National Laboratory, with members of the experimental team responsible for the co-discovery of the J/psi particle in 1976, which earned Ting a share of the 1976 Nobel Prize in Physics (Photograph courtesy of Brookhaven National Laboratory).

teractions. The advent of the quark model in 1964, which would keep quarks and gluons locked up inside nucleons, offered a way around the objection Pauli had raised regarding the conveyors of the strong force. A possible answer lay in the quantum mechanism of "confinement." Although gluons are indeed massless bosons that carry a kind of charge (color charge), they are forever trapped inside nucleons—and, indeed, trapped inside hadrons in general—so they will never be observed moving freely in nature. That explained why gluons had not been seen up to that time and why they were much, much harder to see than another kind of massless particle, the photon (which, ironically, happen to be the only thing that our eyes can see).

While that argument addressed concerns about the strong force bosons (i.e., gluons), the weak force bosons were still problematic. If the weak force were mediated by massless bosons, as prescribed by Yang-Mills theory, it should be a long-range force like electromagnetism—a force of essentially infinite range that is mediated by massless photons. But that was not observed. In fact, experiments had shown just the opposite: The weak force had the shortest range of any of the known forces, implying that the carrier particles associated with it must be extremely massive. Generally speaking, the more massive a particle is, the shorter its mean lifetime. Massive bosons that conveyed the weak force would not stick around very long, meaning they could not travel far before they decayed into lighter particles—a fact that would limit the range of this force to very short distances, just as had been observed.

Theory and experiment were thus in conflict, but a way to heal that rift was found in 1964, which turned out to be an extraordinarily eventful year in particle physics. A new idea was introduced, showing how the weak bosons could have started off without mass and acquired it later.

That notion, subsequently called the "Higgs mechanism," will be the focus of the next chapter. Rather than delving into that topic here, we'll instead jump ahead a couple of years to 1967, when Sheldon Glashow, Abdus Salam, and Steven Weinberg used this radical idea, the Higgs mechanism, in their effort to write a major piece of the "Standard Model"—an all-embracing theory (to be taken up later in this chapter) that attempts to describe the behavior of every known particle. As part of their critical contribution to that effort, Glashow, Salam, and Weinberg brought about major strides in our understanding of the weak force, of the specific particles (or bosons) that confer it, and of a hitherto unknown link it held with another basic force of nature, electromagnetism.

The three physicists built on James Clerk Maxwell's insight from a century earlier, which showed that electricity, magnetism, and light were manifestations of the same underlying phenomenon—a unified force called electromagnetism. The 20th-century theorists showed, in turn, that even though the electromagnetic force (associated with electricity, magnetism, and light) and the weak force (associated with radioactive decay) appear to be quite distinct in our present-day world, they are in essence part of the same basic interaction. At the high energies and temperatures characteristic of the early universe (up until about a hundredth of a billionth of a second after the Big Bang), the electromagnetic and weak forces were unified as a single, "electroweak" force. One could not be differentiated from the other because they behaved exactly the same.

Since that pivotal moment in the universe's history, however, the two forces have parted company. The electromagnetic force is transmitted by massless, electrically neutral photons, whereas the weak force is transmitted by three kinds of massive particles called intermediate vector bosons—the electrically positive W+, the electrically negative W-, and the electrically neutral Z.

Electroweak theory, however, initially suffered from a not-so-trivial problem: When you used it to do calculations, infinities kept popping up, which created a rather considerable nuisance and called some of the theory's more appealing aspects into question. To be truly useful, the electroweak interaction had to be "renormalizable," meaning that it was formulated in such a way as to keep the infinities at bay. In the late 1960s, the prevailing opinion was that electroweak theory, along with the Higgs mechanism it had recently co-opted, was not renormalizable. For that reason, many physicists did not put much stock in this new idea. It was difficult, for example, to calculate detailed properties of the newly proposed W and Z particles without getting unreasonable results.

In 1971, Gerard 't Hooft, a graduate student at the University of Utrecht, published two papers, which overcame the general skepticism by showing how the electroweak interaction could be renormalized. In 1972, 't Hooft published several more papers on the subject with Martinus Veltman, his graduate advisor who had initiated research in this direction. Thanks to the work of 't Hooft and Veltman, the electroweak theory now stood on much firmer ground and could be used to make precise predictions regarding the mass and other features of the W and Z particles.

Glashow, Salam, and Weinberg won the 1979 Nobel Prize for developing electroweak theory in the first place, even though the W and Z particles had not yet been discovered. 't Hooft and Veltman received their Nobel twenty years later, in 1999, and by that time W's and Z's had been produced in vast quantities. The person largely responsible for making the latter happen was the Italian physicist Carlo Rubbia, who had been based at both Harvard and CERN.

Rubbia—along with David Cline, who was then at the University of Wisconsin, and Peter McIntyre, who was then at Harvard—came up with a way of generating these particles by colliding beams of matter and antimatter. Their proposal got a cool reception at first, but Rubbia kept pushing. [38] He took the idea to CERN, where he suggested (in his characteristically forceful manner) that the existing Super Proton Synchrotron (SPS)—which then fired protons into a sta-

tionary target—should be converted into a machine that created head-on collisions between counter-rotating beams of protons and antiprotons. The annihilation of protons and antiprotons would yield far more energy than was possible in
the machine's former configuration. The prodigious energy release, Rubbia
claimed, would give rise to a swarm of new particles, with the W and Z almost
surely to be found among them.

Ultimately, his arguments were persuasive, and in June 1978 CERN approved the proposal to revamp the SPS.[39] Less than five years later, in January
1983, the W+ and W- particles were discovered; the heavier Z particle was discovered several months later. As expected, these particles were quite heavy—
about 80.4 and 91.2 GeV, respectively—and had brief lifetimes of only about
10^{-25} seconds.[40] The properties of these W's and Z's, including their masses,
closely conformed to the predictions of electroweak theory. Rubbia and Simon
Van der Meer, the engineer who worked with Rubbia to reconfigure the SPS, won
the Nobel Prize a year later.

Meanwhile, CERN was already moving ahead with its plans for the Large
Electron Positron (LEP) collider, which would accelerate electrons and positrons
in opposite directions around a 27-kilometer ring. A prime objective of this facility, which was (and still is) the largest electron-positron accelerator ever built,
was to produce W's and Z's in copious amounts so that their characteristics could
be pored over in exquisite detail. Prior to the construction of the LEP, plans were
also being laid to eventually succeed it with a proton collider that would make
use of the same 27-kilometer tunnel.

1983 was noteworthy for another reason in the realm of accelerator physics.
That was the year Fermilab commissioned the Tevatron, which stood as the
world's most powerful accelerator for the next 15 years, so named because it
would bring high-energy physics into the realm of the tera-electron-volt or TeV,
aiming to achieve 2 TeV in proton-antiproton collisions. The Tevatron was the
first large accelerator to use superconducting magnets. It had 1,000 such magnets
whose current-carrying cables were cooled to about 5 degrees Kelvin—a temperature at which the cables can transport large amounts of current without electrical resistance or the loss of energy. Tevatron's superconducting magnets could
produce stronger magnetic fields than conventional magnets—a capability that
was used to bend particle beams around a 6.5-kilometer circle and accelerate
those particles to higher energies.

The Tevatron was the site of a number of key findings, including the aforementioned 1995 discovery of the top quark—the last of the six predicted quarks
to be found. In 2000, a team of researchers at the Tevatron obtained the first direct evidence of the tau neutrino, which is regarded as a relative of the electron—
and may be the last of that family to be discovered as well.

In the century or so between Thomson's identification of the electron as a subatomic particle in 1897 and the discovery of the tau neutrino in the year 2000, experimental physicists had clearly been productive: All told, they had caught glimpses of approximately 200 particles—17 of which (including the most recent entries, the top quark and tau neutrino) are considered to be the fundamental building blocks of nature.

Particle physics, of course, is about more than finding new particles. While discoveries like that are absolutely essential to the process, theorists are also needed to guide the experiments and make sense of the results. And the theorists in this case had been anything but idle. They had spent the last several decades developing a broad framework, which describes all the known particles in the universe as well as the nongravitational forces—strong, weak, and electromagnetic—that act on them. (Gravity, which is too feeble to play a meaningful role in most particle interactions, is treated by a separate theory, general relativity, which Einstein introduced in 1915.) This set of equations or mathematical superstructure, which describes the elementary particles and the relevant forces, is known as the Standard Model. It was mostly finished in the 1970s, and experimentalists have been verifying aspects of it ever since. "Everything that happens in our world (except for the effects of gravity) results from Standard Model particles interacting according to its rules and equations," explains Gordon Kane, a particle theorist at the University of Michigan. [41]

It's somewhat curious that a theory so grand in scope and inspiring in its predictive power ended up with a title that not only sounds mundane but is also somewhat misleading. "The Standard Model is the wrong name," commented Nobel laureate Gerard 't Hooft (at the February 2014 symposium held in Beijing to promote the Great Collider). "It was originally thought of as something to try out [hence the term 'model']," said 't Hooft, who was himself an important contributor to that theory. "But this thing that people wrote down on the back of an envelope turned out to be very precisely correct. It describes everything we see, and all the tests have passed with flying colors. The Standard Model turned into the *standard theory* of elementary particles." [42]

The Standard Model is a crowning example of a quantum field theory—a mathematical formulation that weaves together the principles of quantum mechanics and special relativity while describing forces and particles in terms of fields. At its simplest level, this model is like the periodic table of elements in chemistry, although the "elements" in this case refer to particles rather than types of atoms or chemical species. Every known particle of the Standard Model falls into one of two categories—fermions, of which matter is made, and bosons, which transmit the forces that bind matter together and interact with fermions in other ways. Fermions include all the quarks, which experience the strong force, and all the leptons—lighter particles, such as the electron, that do not experience the strong force and can move freely through space, unlike quarks. Fermions also include composite particles made from some combination of quarks and leptons. Protons and neutrons are classified as *baryons* (derived from the Greek word for heavy)—which are particles that contain three quarks. Pions are classified as *mesons* (derived from the Greek word for middleweight) because they are somewhat lighter, consisting of a quark and an antiquark.

As discussed previously, physicists have identified six different kinds (or "flavors") of quarks, and do not at the moment have convincing reasons to believe there are any more to be found. They have also identified six different kinds of leptons, and every quark and lepton has a corresponding antiparticle.

The electron was the first lepton to be identified. The muon, discovered four decades later, is a heavier relative of the electron. The muon is, in fact, almost identical to the electron in every respect, apart from being about 200 times more massive. The tau is more than 3,500 times more massive than its relative, the electron. The lepton class also includes three varieties of electrically neutral particles called neutrinos.

The existence of neutrinos was postulated in 1930 by Pauli, who observed that during the process of radioactive beta decay—in which a proton and neutron transform into one another, accompanied by the emission of an electron—some of the original energy is unaccounted for and appears to have gone missing. That posed a dilemma: either you accept a violation of the conservation of energy or you propose that there must be an additional particle—one that is small, of neutral charge, and otherwise inconspicuous—to carry the missing energy away. Pauli chose the later option, and a few years later Enrico Fermi dubbed the hypothetical particle the "neutrino"—Italian for "little neutral one." In a letter to a group of nuclear physicists, Pauli confessed that he had taken a "desperate" step. "I have done something very bad today by proposing a particle that cannot be detected. It is something that no theorist should ever do." [43]

It took a couple of decades but validation of Pauli's "desperate remedy" finally came. In 1956, Clyde Cowan and Frederick Reines discovered the first particle of this variety, the so-called electron neutrino, in experiments conducted near a U.S. government nuclear reactor complex in South Carolina. The muon neutrino was discovered in 1962 at Brookhaven's AGS accelerator, and the tau neutrino, as noted above, was found at Fermilab in 2000. [44]

The Standard Model divides these six quarks and six leptons into three families or generations of particles (*See Table 1*). Each family consists of two quarks, an electron or one of its relatives (a muon or tau), and one of the three varieties of neutrinos. The particles in the first family, being the most stable and least massive, are the most familiar ones. In fact, every bit of matter we encounter in the world around us is made up of just three things—up quarks, down quarks, and electrons—which are all members of the first family. A proton, as indicated before, is composed of two up quarks and one down quark, whereas a neutron is composed of two down quarks and an up. The atoms and molecules out of which the human body and the entire Earth are composed are, in turn, composed of protons, neutrons, and electrons.

Particles in the second family are more massive and less stable than those in the first, and particles in the third family are even more massive and less stable still. Members of all three of these families, and combinations thereof, collectively make up every object ever seen in nature, including particles that materialize for just a fleeting moment in high-energy accelerators and disappear just as quickly.

The fact that nature seems to have three families of particles rather than one—or, say, four—is a great mystery that most physicists had not anticipated. The Standard Model itself offers no explanation as to why this is the case, although someday perhaps a deeper theory will do that.

Table 1. The Three Families of Fundamental Particles

1st Family	2nd Family	3rd Family
up quark	charm quark	top quark
down quark	strange quark	bottom quark
electron	muon	tau
electron neutrino	muon neutrino	tau neutrino

The forces of nature are mediated by force-carrying particles called bosons. The electromagnetic force is mediated by the photon—light of varying energies that behaves like a particle on subatomic scales. Two particles interact electromagnetically by sending a stream of photons back and forth between them. The three W and Z bosons transmit the weak force, which is responsible for the radioactive decay of atomic nuclei and other processes. Gluons, carriers of the strong force, come in eight varieties. They bind quarks closely together in protons, neutrons, and mesons, and keep protons and neutrons tightly packed inside the nucleus.

Putting the quarks and leptons (and antiquarks and antileptons) together with the bosons gives physicists all the raw ingredients they need to provide a complete description of the particles and forces (with the exception of gravity) that both populate and shape our universe. The Standard Model, which accomplishes all this, is still far from perfect. There is no apparent way for it to incorporate gravity, for example. Nor can it explain why particles have the masses they do. Despite these limitations, the theory has been a resounding triumph, accurately describing physical phenomena over a vast range of energy scales.

"The Standard Model of particle physics is the single most successful theory in the history of science," claims Fermilab physicist Dan Hooper. "In the decades since its conception, not a single one of the Standard Model's many predictions has ever been shown to be incorrect."[45] The model can accurately predict the rates at which certain particles—like W and Z particles and top quarks—will be produced at accelerators, and how quickly (and into which products) they will decay. It can predict the magnetic response of an electron to an accuracy of twelve decimals place and the response of a muon to about eight decimal places.[46]

The comprehensive sweep of this theory has engendered mixed feelings among some physicists, who have wondered about where we might go from there. Henry Kendall, after winning the Nobel Prize in 1990 with his colleagues Jerome Friedman and Richard Taylor, has expressed such qualms. Their research, which affirmed the quark model as well as the theory of the strong force—quantum chromodynamics—helped pave the way to the Standard Model. "And that [theory] is such an astonishing success that it's depressing," Kendall said. "There's nothing left to do."[47]

Although there was much truth to Kendall's observation, which expressed a sentiment that others certainly shared, it was, nevertheless, something of an over-statement. For at that particular moment in history and even a decade later—following the 2000 discovery of the tau neutrino, after which all the quarks and leptons had been identified—there was still something missing from the theory. And that "something" may, in some ways, have been the most important thing of all. For without it, the Standard Model—the culmination of a century of stunning progress in high-energy physics—would have come crashing down.

Chapter 2

Chasing the Higgs

AT THE TURN OF THE 21ST CENTURY, where our story left off, all the particles of the Standard Model had been discovered—except for one, the Higgs boson. That was a rather sizeable omission, given that the Higgs has been called the lynchpin of the entire theory or, as Nobel laureate Gerard 't Hooft has put it, "the final ingredient."[1] And in some ways this missing particle may be the most critical ingredient of all—perhaps not the plums of J. J. Thomson's metaphorical plum pudding, but more like the starch that enables the liquid base to firm up and become a pudding in the first place. For according to theories formulated in the early 1960s by Peter Higgs and several other physicists, there is an invisible field associated with the Higgs particle that suffuses all space with a form of energy, even pervading regions that are devoid of matter and might otherwise appear empty.

Most of the known elementary particles, including electrons, acquire mass through their interactions with this invisible field—the Higgs field—which they encounter at every turn and are constantly jostling against. The mass accrued in this way has enabled atoms and molecules to form—and from them larger entities like stars and planets. Without the Higgs field conferring mass to them, elementary particles would zoom around the cosmos at the speed of light, having little if anything to do with each other. These particles, which we call nature's building blocks, would not sit still long enough to build anything. The universe, consequently, would be a very different place—and one that humans (and other life forms) would not be around to enjoy or to devise intricate theories about. Although energy and stray particles would abound, there would be no ordinary matter or chemistry or biology. There would, quite literally, be little of substance to talk about and no one to do the talking.

Of course, today's universe is not stuck in this barren, matter-less state, and a theory drawn up in 1964 helped explain why. This theory specifically helped answer the question of why some particles have mass while others, such as photons, do not. Building upon earlier work by Philip Anderson, Jeffrey Goldstone, and Yoichiro Nambu, six physicists published papers during a three-month span in that year, postulating the existence of an almost magical field, which fills space with something akin to cosmic molasses. This pseudo-molasses backdrop is not felt by massless particles like photons, which are impervious to the field, passing through it unperturbed. But all massive particles experience a kind of drag as they move through the unseen medium. This resistance to motion, a property known as inertia, is intimately tied to mass: The greater an object's mass, the more it resists changes in its motion or acceleration. Particles pick up more or less mass depending on the extent to which they interact with the Higgs field. Particles that travel quickly through the field, interacting with it minimally, take on less mass; those that keep bumping against the ubiquitous Higgs field move through it more slowly and hence take on more mass.

The above story is, admittedly, far too simplistic. It suggests that particles fritter away energy as they move through space, interacting with the Higgs field along the way, just as a projectile would lose its "oomph" after plummeting into a vat of molasses. This is not the way that physicists really picture it. Interactions with the Higgs field would not siphon off a particle's kinetic energy. But these unrelenting bombardments would, nevertheless, slow it down. In fact, the constant buffeting by the Higgs field would keep a particle from moving at the speed of light, and we know from special relativity that particles that cannot move at the speed of light must have mass.

So what the Higgs field does, in a sense, is to put up all sorts of roadblocks or "speed bumps" out in space, which hinder a particle's motion, thereby localizing its energy. And the confinement or localization of energy around a given area is one of the things that we mean by "mass."

Another way to think of this, suggests Harvard physicist Matt Strassler, "is that the Higgs field changes the environment of the whole universe so that certain particles—those that interact directly with the field—take on different properties, and behave differently, than they did before."[2]

A couple of decades ago, University College London physicist David Miller offered the following analogy to a British science minister in an attempt to explain the Higgs mechanism.[3] Consider a cocktail party in which people are distributed uniformly throughout the room. An unknown person dressed in ordinary attire might walk across the floor at a rapid pace, without attracting any notice or exchanging words with anyone—analogous to a massless photon sailing through the Higgs field unimpeded. On the other hand, a celebrity—a big-name screen actress, rock diva, or headline-grabbing politician—would attract a crowd as she made her way across the room and would therefore have to work harder to reach her destination (such as the bar or the hors d'oeuvres tray). Even though her pace in traversing the room is slower and more labored, her movements are still associated with greater momentum, and hence greater effective mass, because of the entourage that is constantly following her around and matching her every move. Stopping is therefore harder, but once she does stop somewhere—to sign autographs, for instance, or to answer questions or pose for a photograph—it will also be harder for her to get going again because of the cluster of people surrounding her. This celebrity, who has continuous encounters with others in the room, is analogous to a massive particle, such as the top quark, which interacts very strongly with the Higgs field and thus has to push more vigorously to get moving and to keep moving. [4]

The first paper about this hypothetical field—written by François Englert and Robert Brout—came out in August 1964. Peter Higgs published two solo papers on the same theme in September and October of that year, followed by a joint paper in November by Gerald Guralnik, Carl Hagen, and Thomas Kibble. Yoichiro Nambu, who served as referee for Higgs's second paper, encouraged the author to say more about the physical implications of his theory. Higgs added a brief section, which explained that an excitation of what came to be known as the Higgs field—jiggling it through the infusion of energy in just the right way—would unleash a new particle, just as exciting the surface of a calm sea (by a burst of wind or dropping a heavy object) will create a ripple or wave. [5]

Peter Higgs secured his place in history by being the first person to emphasize the existence of this extra particle, which was eventually named after him, as was the field associated with it. By calling attention to the new particle, Higgs gave experimentalists a clear-cut goal, while providing them with a concrete object to hunt for. Even though the Higgs field offered a mechanism for solving the crisis in theory that was behind all this fuss, no one has figured out a way to detect that field directly. So what one does, instead, is to look for the Higgs boson, which is a tangible manifestation of the field that shares its name.

This is the strategy that Brout, Englert, Guralnik, Hagen, Higgs, and Kibble cooked up. They were motivated by the seeming paradox that arose during the construction of the Standard Model: The theory demanded a broad kind of symmetry—and was indeed built upon it—while that symmetry seemed to demand, in turn, the existence of certain particles (massless charged bosons) that do not, so far as we know, appear in nature. Each force of the Standard Model—the strong, electromagnetic, weak, and combined electroweak—has an associated symmetry of a sweeping nature called gauge symmetry. These underlying gauge symmetries, which lie at the heart of Yang-Mills theory (as discussed in Chapter 1), are essential to the Standard Model. Without them, the theory would make nonsensical predictions regarding, for instance, processes that have probabilities greater than 100 percent—something that obviously cannot occur.

A system, force, or field endowed with gauge symmetry can undergo various transformations, including translations or rotations, without affecting any of the essential physics. Forces in the Standard Model can also withstand another kind of transformation: Particles that carry the forces can be interchanged at will, and no one could tell the difference. This is considered a highly symmetrical arrangement because one force carrier can be substituted for another without any discernible consequence; everything would still look the same after the switch, which is precisely what physicists mean by "symmetry." What this also implies is that at sufficiently high energies and temperatures, when the electromagnetic and weak forces are combined as one, the electroweak force carriers—photons and W and Z particles—are identical to each other in terms of mass. And since photons are massless, W and Z particles must have been massless too in the era of electroweak unification. But therein lies the rub: Physicists did not (and still do not) believe that massless, electrically charged particles exist; the notion did not make sense on theoretical grounds nor did it accord with the experimental evidence. Wolfgang Pauli pressed Chen Ning Yang on this point in 1954 (as discussed in Chapter 1). The beautiful mathematics of Yang-Mills theory was put on hold for about a decade until the issue could be sorted out and the theory triumphantly resurrected.

One could not simply assign masses to the W and Z particles in order to bring Yang-Mills theory into better agreement with empirical indications at that time, because such a change would have ruined the mathematical symmetry that gave rise to the weak force in the first place. Nevertheless, a way out of this dilemma was found by assuming that the universe's first particles were indeed born massless, but that the symmetry that made the elementary particles indistinguishable by mass was short-lived. Something happened within about a 100 billionth, or 10^{-11}, of a second of the Big Bang (as discussed in Chapter 1): The electromagnetic and weak forces went their separate ways, breaking the "symmetry of masses." And the universe, which had been cooling steadily since the Big Bang, underwent a "phase change"—a transformation similar to what happens when water turns into ice.

A change of this sort, going from a warmer to a cooler state, is typically accompanied by a decrease in symmetry. "Although you might intuitively think that something more ordered, like ice, is more symmetric, quite the opposite is true," says Columbia physicist Brian Greene. "Something is more symmetric if it can be subjected to more transformations, such as rotations, while its appearance remains unchanged." Ice's crystalline structure makes it invariant to very limited types of rotations of a specific magnitude (such as 60 degrees) and around specific axes. Because molecules of liquid water are laid out more randomly, they can be rotated around any axis, and by any angle, without changing the liquid as a whole in a discernible way.

Symmetry is increased even further when water boils and is converted to a gaseous form, because molecules in a gas are even less aligned than those in a liquid. Rotations of a gas, therefore, leave a system closer to its original state, which is a mark of greater symmetry. Going from a gas to a liquid, conversely, is accompanied by a decrease in symmetry, and this may be the best way of thinking about the transition that occurred within the universe's first fraction of a second. The phase change would have been set off as the universe expanded and cooled, until the cosmic temperature dropped below the "boiling point." The Higgs field—which had previously been fluctuating, having an average value, or energy, of zero—condensed to form a kind of "ocean," as Greene describes it, with a uniform, nonzero value that spread out to fill every inch of space.[6]

Turning up the Higgs field in this way meant that the symmetry, which had until then made the electroweak force mediators massless, was suddenly and spontaneously broken. From that point forward, the electromagnetic and weak forces were no longer one and the same. While they had once been of equivalent strength, the weak force started living up to its name and henceforth appeared to be weaker than electromagnetism in the so-called "low-energy" conditions that set in after this phase transition. In keeping with that change in status, the carriers of those forces—photons and W and Z bosons—were no longer interchangeable either. The authors of this idea—Brout, Englert, Guralnik, Hagen, Higgs, and Kibble—explained how all that could have transpired. Their solution, in essence, was that the gauge equations at the heart of electroweak theory were correct; it's just that the universe itself was different from what had been originally supposed. It wasn't empty, for one thing.

Other physicists, including Steven Weinberg, drew on the prior work of Hideki Yukawa to suggest that other elementary particles in the Standard Model— quarks and charged leptons, but not massless photons and gluons—would also interact, to varying degrees, with the omnipresent Higgs field, gaining different amounts of mass in the process. Before these theorists had made such a leap, they faced a problem—which was that the equations governing the weak nuclear force made bizarre, unsound predictions when masses were indiscriminately given to W and Z bosons. (This issue, which relates to so-called "W-W scattering," is taken up in Chapter 5). However, with the adoption of the proposed symmetry-breaking mechanism, the equations that describe the weak and electroweak interactions could now accommodate both massless force carriers (like photons) and massive force carriers (like W and Z particles), as has been confirmed by experiment.

The symmetry inherent in nature's laws did not disappear. It's still there, part of the basic fabric, and could be restored in full force simply by raising the universe's temperature and "evaporating" the Higgs field. But the actual state of the world—depending on the prevailing temperature and other factors—masks the original symmetry, which remains hidden from view.

For example, in some cases it may be difficult to discern rotational symmetry—a property shared by circles and spheres in which rotations around their centers leave them looking unchanged. In this situation, all directions and orientations in space are equivalent, with none favored over another. But circumstances can arise in which a preferred direction is arbitrarily picked out—a choice that is not dictated by logic but is instead the consequence of a random act. "Think of a pencil that originally stood on end and then falls down and chooses one particular direction," suggests Harvard physicist Lisa Randall, "All of the directions around the pencil were the same when it was upright, but the symmetry is broken once the pencil falls." That's true because a pencil cannot fall in all directions at once; it has to pick just one direction in which to topple. "The horizontal pencil thereby spontaneously breaks the rotational symmetry that the upright pencil preserved," says Randall. Similarly, she adds, "the Higgs field, which permeates the universe in a way that is not symmetric ... spontaneously breaks weak force symmetry."[7] The net result is that this mechanism allows particles, which came into the universe massless, to gain some heft.

But which particles, exactly, does the Higgs field confer mass to? The W and Z bosons, according to the Standard Model, owe their mass to interactions with that field. That should come as no surprise since part of the impetus behind the Englert-Brout-Higgs-Guralnik-Hagen-Kibble mechanism was to make these particles weighty. In addition, all the charged leptons—the electron, muon, and tau—get their mass from the Higgs field, whereas massless particles like photons and gluons don't interact with that field at all. (The question of whether, and the extent to which, the lightest particles—the uncharged leptons called neutrinos—get their mass from the Higgs field is not well understood, and there's a divergence of opinion on this issue. Physicists are not entirely sure how neutrinos come by their modest mass, although the topic will be discussed further in Chapter 4.)

"When we say that the mass of the electron is created by interactions with the Higgs field," explains Fermilab director Nigel Lockyer, "we can think of this as the Higgs field rapidly changing a left-handed electron into a right-handed electron and vice versa." Labeling an electron left-handed means that it spins in a direction that is opposite to its motion or momentum. Conversely, a right-handed electron's spin is aligned with its motion. (The spin direction here is in-

CHAPTER 2: CHASING THE HIGGS

dicated by the right-hand rule: if the particle is spinning in the direction of the curled fingers of the right hand, then the spin direction is said to be aligned with that of the extended right thumb.) "This switching back and forth [takes] energy," Lockyer says, "and through $E=mc^2$, energy is mass." The same mechanism, involving the rapid oscillation between right and left-handed forms, is thought to give mass to quarks, muons, and other particles. "A heavier particle like the top quark would experience this flipping at a much higher frequency than a lighter particle like the electron," he adds, with its energy or mass being directly proportional to its frequency. [8]

The fact that electrons are massive is of more than passing interest to us. If electrons were lighter, atoms and molecules would be larger and much less stable. If electrons were massless, the size of an atom would effectively be infinite, meaning that electrons would not bind with protons and neutrons to form atoms. And without atoms or molecules, the world that we see today—including us—would not be here.

Quarks, as alluded to above, get their mass from the Higgs interaction as well. Given that protons and neutrons are nearly 2,000 times more massive than electrons, and that protons and neutrons are themselves made up of quarks, one might assume that the mass of ordinary matter is mostly due to quarks. But that's not the case. Although the Higgs field gives mass to elementary particles like quarks and electrons, explains Brian Greene, "when these particles combine into composite particles like protons, neutrons, and atoms, other (well understood) sources of mass come into play." [9] The total energy of a composite particle consists of the rest mass of its constituents and their kinetic energy of motion, plus the combined energy of the interactions between these constituents.

That helps explain why quarks account for only about one percent of the proton's mass. [10] The bulk of a proton's mass comes from the kinetic energy of its quarks and gluons, which move around inside nucleons at nearly the speed of light, and from the binding energy of the strong force that holds those quarks together. However, the inside of a proton is far more complicated and chaotic than a mere collection of three constituent quarks and associated gluons. It is literally filled with what are sometimes called "virtual particles"—virtual gluons and pairs of virtual quarks and antiquarks—that are constantly (and fleetingly) flickering in and out of existence. This phenomena is a consequence of a standard tenet of quantum mechanics, the uncertainty principle, which holds that a small particle never sits completely still, nor does the field that gives rise to it. The particle and its associated field are constantly jiggling around, and the resultant jiggles or fluctuations can create ripples that, temporarily, generate even more (so-called virtual) particles. The energy of the fluctuations responsible for these nonstop comings and goings contributes to the total mass of the proton. And the same general picture holds for the neutron as well.

As a consequence, explains Juan Maldacena of Princeton's Institute for Advanced Study, even though "the Higgs gives mass to most elementary particles, most of the mass [around us] does not come from the Higgs field." Instead, it mainly comes from the mass of protons and neutrons. [11]

It is for this reason, among others, that most physicists, including Gerard 't Hooft, resist the popular designation of the Higgs boson as the God particle. "We don't believe in a God particle," 't Hooft affirms. "We just believe in systematic mathematics and physics and the laws we are trying to discover." [12]

So, how did "systematic physics" triumph in this instance, by finding and identifying the Higgs boson? For nearly half a century, the Higgs boson was merely a hypothetical entity—albeit one that held a critical place in theory. Physicists long believed that the best way to discover such a particle would be to slam two protons or a proton and antiproton together, each moving in opposite directions at nearly the speed of light. "We do all of this work in order to put a little ripple in the Higgs field that fills the universe," explains the theorist Nima Arkani-Hamed. "Mostly it's like the surface of a quiet lake where nothing is going on. But if you bang it hard enough, you can put a tiny ripple on the surface of that lake. And the smallest possible ripple you can put, consistent with quantum mechanics, is called the Higgs particle." [13]

One should not picture this lake or—to return to our previous analogy—"ocean" as being filled with Higgs particles in the same way that oceans on Earth are filled with water molecules. The omnipresent Higgs field (or Higgs ocean) is generally devoid of such particles, except when the precise input of energy—through collisions and other means—creates tiny ripples on its surface. In quantum theory, all elementary particles can be thought of as ripples in a corresponding quantum field. Ripples in the Higgs field, not surprisingly, correspond to Higgs particles, but these particles are short-lived, lasting just a billionth of a trillionth of a second before they dissipate, and calmness is restored to the ocean's surface.

The key challenge for experimentalists was to build an accelerator that could produce collisions violent enough to create ripples that would, in turn, trigger the release of Higgs bosons. There were hopes that Fermilab's Tevatron, which was the world's most powerful particle accelerator for a quarter of a century, might discover the Higgs. The physicist Robert Wilson, who served as Fermilab's director, made the case for this machine—and the predecessors from which it evolved—at a 1969 U.S. Congressional hearing prior to the start of construction. In response to questions posed by Rhode Island Senator John Pastore about how the accelerator might enhance national security, Wilson maintained that the new accelerator would have "nothing to do directly with defending our country except to make it worth defending." [14]

Proving its value many times over during its 1983 to 2011 run, the Tevatron was responsible for the discovery of the top quark, the tau neutrino, the five so-called "B baryons" (all of which contain the bottom quark), the B_c meson (which consists of a bottom quark and charm quark), and other particles, but fell just short of the mark in the hunt for the Higgs boson. The Tevatron clearly made Higgs particles; it just didn't turn them out in sufficient quantities for physicists to be sure of what they had. While researchers at Fermilab saw strong hints of a particle in the mass range of 115 to 135 GeV, that effort only merited a statistical significance of "3 sigma," meaning the odds of their findings being due to pure chance were about one in 550.[15] Although that was an encouraging result, discoveries in physics must meet a much higher standard of 5 sigma, where the odds of a finding being a statistical fluke—a consequence of background fluctuations or random "noise"—are less than one in 3.5 million. The goal, putting this in other terms, was to detect a signal that we could be virtually certain came from something new, namely the Higgs itself, rather than from Standard Model particles we were already acquainted with. The Tevatron greatly narrowed the energy range in which physicists should look for the Higgs, but it simply could not produce collisions at a high enough rate, and at high enough energies, to reach the discovery threshold.

The Superconducting Super Collider (SSC), which was slated to be built south of Dallas, Texas, would have sent protons hurling around a 87-kilometer-long (54-mile-long) oval-shaped tunnel, producing collisions that reached energies of 40 TeV—or 20 times more powerful than the Tevatron—a level, it was felt, that would have almost guaranteed it to find the Higgs. Planning for this mammoth project began in 1982, and seven years later a House-Senate Committee allocated money for construction, which began in 1991. However, the SSC project suffered from bureaucratic red tape, cost overruns, persistent questions about management, and a constantly shifting political climate (including the inauguration of a new president in 1993).

Eloquent arguments in support of basic research in particle physics, such as those made by Robert Wilson two decades earlier, held little sway in the early 1990s. The House of Representatives voted to cut off the SSC's funding in June 1992, although the Senate restored funding two months later. In October 1993, Congress reversed itself again, killing the project outright after $2 billion had been spent[16] and almost 24 kilometers (15 miles) of tunnels dug into the vast chalk deposits (more specifically limestone, sandstone, and shale) that underlie much of central and southeastern Texas.[17] "We went at this with the painfully derived support of three administrations and got it 20 percent finished," said Nobel laureate Leon Lederman, venting the frustration shared by many of his colleagues. "And then Congress, in its infinite wisdom, said, Oops, no."[18]

Sherwood Boehlert, a former Republican congressman from New York, gave a sense of the mindset that the physicists were up against: "I doubt anyone believes that the most pressing issues facing the nation include an insufficient understanding of the origins of the universe." [19] Pervasive sentiments along those lines helped bring about the cancellation of the SSC—a decision that dealt a blow to the U.S. high-energy physics community from which it has not yet recovered—and from which it may never recover.

Fortunately, a backup plan was in the works, albeit on the other side of the Atlantic. CERN's Large-Electron-Positron (LEP) Collider—the largest facility of its kind ever built—had turned on in 1989 and was in full operation at that time. Years before construction began in 1983 on that accelerator and its 27-kilometer-circumference tunnel, a decision had been reached to eventually install a proton collider there that would utilize the same tunnel. This machine, which came to be known as the Large Hadron Collider (LHC), "had a difficult birth," according to Lyndon Evans, the British physicist who was charged with making the proposed collider a reality. "The approval of the SSC ... in 1987 threw the whole project into doubt. The SSC, with a center-of-mass energy of 40 TeV, was almost three times more powerful than what could ever be built at CERN." The only thing that kept the project alive, added Evans, was "the resilience and conviction of Carlo Rubbia." [20]

Rubbia pushed forcefully for the LHC at a time when Europe was experiencing fiscal challenges of its own, and it was fortunate for the field of high-energy physics that he managed to exert his will. For when the SSC was scrubbed in 1993 and the Tevatron began winding down its operations more than a decade and a half later, the physics community had somewhere to turn—and that was towards Geneva.

The LEP shut down in 2000 to make room for the LHC. After construction was finished in the summer of 2008, with about $8 billion spent up to that time, [21] the collider circulated its first beams on September 10—an occasion that was widely celebrated. However, LHC physicists suffered a setback just nine days later when an electrical malfunction caused a helium leak inside the machine's tunnels. The collider was shut down, and it took more than a year to fix the problem, which involved the removal, repair, and replacement of 53 magnets in a three-kilometer stretch of the tunnel. [22] Scientists and engineers got beams circulating, once again, in November 2009, and the accelerator was back on track in March 2010, when it achieved its first proton collisions.

Delays of some kind are practically unavoidable when a machine of this scale and technical sophistication is brought online. "The LHC is the most complex scientific instrument ever constructed," claimed Evans, who oversaw its building from design to completion. [23]

Robert Wilson, the first director of the Fermi National Accelerator Laboratory (Fermilab), is shown in 1969 at the groundbreaking ceremony for a machine, which would—after a series of additions and upgrades—eventually become the Tevatron collider. (Photograph courtesy of Fermilab)

The Superconducting Super Collider (SSC)—slated to become an 87-kilometer-long particle accelerator south of Dallas, Texas—was cancelled by the U.S. Congress in 1993 after it was about 20 percent complete. (Photograph courtesy of the U.S. Department of Energy)

The Tevatron particle accelerator, which was more than six kilometers in circumference, as seen at night. The Tevatron stopped operating in 2011. (Photograph courtesy of Fermilab)

The main purpose of this elaborate contraption (far beyond anything that Rube Goldberg likely imagined) is to get beams of protons flowing in opposite directions along a 27-kilometer-long racetrack that straddles the border between Switzerland and France. After accelerating to 99.999999 percent of the speed of light, these protons—and their constituent quarks and gluons—ram into each other, reproducing in their fiery smashup conditions that occurred a tiny fraction of a second after the Big Bang.

Thousands of superconducting magnets of different varieties and sizes guide the beams around the accelerator's circular track. These magnets are cooled to -271.3 Celsius—colder than outer space and just two degrees above absolute zero—to eliminate electrical resistance and associated power losses. The LHC employs some of the most potent magnets ever built, generating magnetic field strengths some 100,000 times more forceful than Earth's. Magnets deflect charged particles like protons from moving along a linear path. The faster the protons are moving, the harder it is to push them onto a fixed, curved trajectory. By incorporating the strongest magnets available, which make up the largest parts of the machine overall, the LHC can bend the fastest moving particles and thus attain the highest energies possible. [24]

When the accelerator was running at full capacity during its first phase of operations—producing "center-of-mass" energies of up to 8 TeV—collisions occurred at the rate of about one billion per second. [25] These collisions take place at four spots along the 27-kilometer course where the proton beams cross, and a detector is installed at each of these intersection points to inspect the carnage. Two of the four detectors, ALICE and LHCb, are not involved in the Higgs search. ALICE studies collisions of lead nuclei and other atomic nuclei, instead of protons, when the accelerator is operating in its "hadron" mode. (Which is why this facility is called the Large Hadron Collider rather than the Large Proton Collider.) LHCb focuses on the decay of bottom quarks.

Generally speaking, the LHC was built to study matter at temperature and energy levels roughly seven times higher than had been reached before. But the LHC's two giant detectors, ATLAS and CMS, were designed with a more specific objective in mind, comments Lisa Randall: "to find [the Higgs] particle, show it did not exist, or demonstrate that a more complicated or subtle model was at work." [26] The goal was not simply to find the particle but rather to study the particle in the hopes of gaining a better understanding of the Higgs field and how it works.

ATLAS and CMS are positioned at opposite ends of LHC's tunnel. For the purposes of cross checking, these devices use separate pieces of gadgetry to analyze the outcomes of proton-proton collisions. ATLAS, short for the unpithily named "A Toroidal LHC Apparatus," has been described by the physicist Monica Dunford, a member of the research team, as "five stories completely filled with microelectronics, all custom-designed and hand-soldered, like a [giant] Swiss watch."[27] More than 3,000 physicists are engaged in the collaboration. Although ATLAS and CMS (the Compact Muon Solenoid) employ somewhat different technology, they share several common features.

Both instruments have four main components: First, there's an inner detector or "tracker," which maps the precise trajectories of charged particles and measures their momenta by picking up the electrical signals generated by these compact, swiftly moving objects. Part of this job is handled by silicon detectors, which use microscopic wires printed on silicon microchips to register the presence of ions. Second, there's an outer spectrometer to identify and measure the momenta of muons. Calorimeters, which generally employ light sensors to measure the energy carried by particles, make up the third key component. Last but not least is a system of magnets, which don't detect particles directly but nevertheless bend charged particles in revealing ways to facilitate momentum measurements: The direction that a charged particle bends in the presence of a magnetic field tells us whether that particle is positively or negatively charged. The extent of the deflection depends on the particle's mass; the lighter the particle, the more it gets deflected.

ATLAS's gargantuan detector, which has been said to resemble "an alien spaceship,"[28] is roughly 45 meters long and more than 25 meters tall, making it about half the size of the Notre Dame Cathedral. Weighing in at about 7,000 tons, ATLAS is as heavy as the Eiffel Tower or one hundred 747 jets.[29]

CMS appears to be somewhat smaller than ATLAS—the volume of a more modest-sized cathedral—but it is about twice as massive, using 12,000 tons of iron alone and incorporating some of the most powerful magnets found anywhere.[30] It also supports an even bigger collaboration than ATLAS, with some 4,000 physicists taking part in the effort.

Given the extraordinary lengths that people have gone to in an attempt to detect the elusive Higgs, and the equally extraordinary technology that has been brought to bear on this problem, it is not surprising that CERN has sometimes referred to this endeavor—and to particle physics in general—as "the unbelievable in pursuit of the unimaginable."[31]

The ultimate target of this "unimaginable pursuit," the Higgs boson, is highly unstable, remaining intact for less than 10^{-21} of a second. That's less time than it takes for light to traverse the length of a single atom.[32]

Owing to the particle's ultra-brief lifetime, researchers cannot hope to glimpse the Higgs itself but can obtain signs of its existence by identifying the lighter particles it decays into—and the still lighter particles that they, subsequently, decay into. Physicists specifically try to tally the energy and momentum of *all* the particles emerging from a proton-proton collision, while analyzing the

The ATLAS detector at the Large Hadron Collider at CERN (Photograph courtesy of CERN)

The CMS detector at the Large Hadron Collider at CERN (Photograph courtesy of CERN)

trajectories they take, in order to determine the mass and other properties of the particle that spawned them.

The most effective way of making a Higgs boson at the LHC to date has been by smashing two gluons together. The particle's subsequent decay can follow a number of different pathways or "channels," some of which occur more frequently than others. The Higgs, for example, can decay into two energetic photons (or so-called "gammas"); or into a bottom quark and its antiquark; or into a tau lepton and its antiparticle; or into a W boson and its antiparticle and two Z bosons, which in turn decay into four charged leptons—an electron, positron, muon, and antimuon. The dominant decay channel for the Higgs is into the bottom quark and bottom antiquark, but these events are less revealing to investigators because it's a million times more likely that the bottom quarks and antiquarks came from the decay of ordinary particles than from a Higgs particle. [33] This constitutes one of the greatest challenges that experimentalists face, alongside the fact that the Higgs decays so quickly it cannot be seen directly. "Most of the time the Higgs decays into something that is indistinctive, so the new thing you are looking for gets drowned out in a background of stuff you already know," explains Matt Strassler. [34] Finding the Higgs particle, he adds, "is in some ways analogous to trying to see a distant star during daylight hours; you're seeking a very dim signal against a very bright background." [35]

One strategy would be to pick a different signal to look for, one that's even dimmer than the first, but whose background is far less bright—like looking for an ultra-faint star at night rather than looking for a brighter one during the day. This approach has paid off at the LHC, where the most fruitful search strategy so far has involved the decay of a Higgs particle into two photons. This process occurs only about 0.2 percent of the time that a Higgs is made, but the decay is rather distinctive and can thus be more readily tied to a Higgs progenitor. [36]

It should be stressed, moreover, that only about *one out of a trillion* proton-proton collisions give rise to a Higgs that is detectable at the LHC—a statistic that helps illustrate the immensity of the challenge that experimentalists face. They not only have to pick out that exceedingly rare event, they also have to avoid becoming swamped by a potentially overwhelming flood of data. With proton collisions occurring at a rate of about a billion per second, data from roughly one thousand of those collisions deemed most promising is saved; the rest is instantly thrown out in an attempt to avoid information overload. [37]

A preliminary step involves relying on theory to figure out what the general background should be—to determine, in other words, the number of collision events of a given type (such as those yielding two photons) that would be expected without any Higgs particles on the scene. A significantly greater number of events of a particular type than expected could provide evidence of the Higgs—the more of these events spotted, the stronger the evidence. A sharp spike in the data at a particular energy would point to the existence of a particle of that same energy or equivalent mass. The caveat here is that physicists need to detect a high enough number of collision events to convince themselves (and others) that they are actually seeing signs of a new particle rather than being fooled by chance fluctuations.

The LHC researchers followed the same general procedure. They examined as many of the collisions that led to the production of two photons as they could get their hands on, calculating the combined energy of the two photons in each collision. Although one would expect a rather smooth distribution of events at different energy levels, due to the random processes that can lead to the production to two photons, the experimentalists looked to see if there was a characteristic bump or excess at a particular energy—one that might correspond to the mass energy of the Higgs particle. They went through the same process for collisions that produce four charged leptons, which have another relatively distinctive signature. "If you see a bump in both plots, and the bumps are at the same mass, that's evidence that you have a new particle decaying both to two photons and four leptons," explains Strassler. [38]

By late June 2012, the ATLAS and CMS team members felt confident they had amassed enough evidence to make a strong, and essentially ironclad, case for the Higgs. A public announcement on this subject at CERN was scheduled for 9 a.m. on July 4—a day picked to coincide not with Independence Day in the United States but rather with the start of the International Conference on High Energy Physics in Melbourne, Australia. Anticipation prior to the CERN press conference was intense, to say the least. Hundreds of physicists and other interested onlookers waited in line hours beforehand, some spending the night outside the main auditorium in the hopes of getting a seat to what was likely to be a momentous occasion. Thousands of others watched the proceedings, which were broadcast live over the internet.

CERN Director-General Rolf Heuer greeted the audience, promising a "special day" highlighted by presentations from both the CMS and ATLAS teams. [39] First up was Joseph Incandela, the CMS spokesperson who thanked the 4,000 members of the collaboration. Incandela quickly got to the point, unveiling data that showed a spike of two photon events at about 125 GeV, as well as a 125 GeV peak from four lepton events. Combining data from these two channels, the result had a 0.6 GeV uncertainty and an overall statistical significance of 5-sigma. [40] The evidence amassed by the CMS team, Incandela said, provided a compelling case for "one of the rarest particles ever made, and that's what we call the Higgs." [41]

The ATLAS spokesperson Fabiola Gianotti took the stage next, discussing the spike that her team detected at about 126 GeV. After pooling data from two photon events with that of four lepton events—"the grand combination," as Gianotti

The physicist Carlo Rubbia in his office at CERN
(Photograph courtesy of CERN)

The physicist Joseph Incandela, former spokesperson for the CMS collaboration, as seen monitoring
results from the CMS control room (Photograph courtesy of CERN)

The physicist Fabiola Gianotti—former spokesperson for the ATLAS collaboration and the future director-general of CERN—as seen standing by the ATLAS detector (Photograph courtesy of CERN)

The physicist Rolf Heuer, Director-General of CERN, announces the discovery of what is believed to be the Higgs boson, at a July 4, 2012 press conference. (Photograph courtesy of CERN)

François Englert, co-winner of the 2013 Nobel Prize in Physics, while touring the Large Hadron Collider in 2007 (Photograph courtesy of CERN)

A 2008 portrait of the physicist Peter Higgs, co-winner of the 2013 Nobel Prize in Physics
(Photograph courtesy of CERN)

put it—the ATLAS group also reached the 5-sigma discovery threshold. In a moment of levity, she expressed gratitude that the new particle's mass was at an energy level that the LHC was able to observe and study. "It is very nice of the Standard Model boson to be at that mass," Gianotti said. "Because of that mass, we can measure it. Thanks, nature." [42]

Having corroborating evidence from two separate experiments, which used different apparatus to analyze the outcomes of proton-proton collisions, helped convince physicists worldwide that a Higgs particle—with a mass somewhere in the vicinity of 125 to 126 GeV—had indeed been found. The crowd broke into a chorus of applause, cheers, and whistles. "As a layman," Heuer called out over the din, "I will now say, I think we have it!" [43]

Four theorists attending the proceedings—Englert, Guralnik, Hagen, and Higgs—who had conceived of the Higgs field in 1964, were given a standing ovation. There'd been a long wait prior to this celebration, as the 48-year lag between prediction and discovery was arguably the lengthiest in the history of particle physics. [44] Upon taking the microphone, Higgs, who was then 83 years old, congratulated the physicists participating in the experiment. "For me, it's really an incredible thing that it happened in my lifetime," he said.

"Not only in your lifetime," added Heuer. "Everyone who was involved in the project, and is involved, should be proud of this day. Enjoy it!" [45]

The "it" to which Heuer referred was indeed a collective triumph. Construction of the LHC and the research that has proceeded there involves the collaboration of more than 10,000 scientists and engineers from more than 100 countries. [46] "If only the United Nations could work like CERN," noted Fermilab theorist Joe Lykken, "the world would be a better place." [47]

Discovering this particle was an extremely difficult thing to do, serving as "a model for how the world can cooperate," Incandela said, with Palestinians and Israelis, Iranians and Iraqis, and Indians and Pakistanis working side by side. "We worked for about twenty years in preparing the machine and its detectors. And when we began, we did not have the technology that we needed in the end to actually do what we did.

"We sometimes are asked," Incandela added, "'Does it really take so many physicists to do one experiment?' And yet if you really know what's involved, it's surprising that it takes so few." [48]

Out of the thousands of physicists who had a hand in finding the Higgs particle, just two—Englert and Higgs—garnered a 2013 Nobel Prize for the efforts on this front. The Nobel Committee limits the number of prizes to three living recipients. Brout, Englert's coauthor, would likely have been in line to receive a prize as well, but he died in 2011—a year before the breakthrough at the LHC.

That discovery is ushering in a new era of physics, leaving many questions to be addressed, because at a basic level the Higgs boson is not only a new particle but also a new kind of particle. "Part of what make the Higgs particle so fascinating," explains Edward Witten, "is that it's the only elementary particle that we know about that doesn't have spin."[49]

"We've seen a whole zoo of particles over the last 60 years or so of particle physics, but we've never seen one like the Higgs before," adds Arkani-Hamed. "Although we have been anticipating the Higgs for decades, that doesn't change the fact that it is profoundly mysterious that it exists."

Every elementary particle has an intrinsic, unvarying property called spin, which is a kind of angular momentum. All bosons have an integer spin, whereas all fermions have a half-integer spin. From a theoretical standpoint, the only spins that a particle is allowed to have are 0, ½, 1, 3/2, and 2. "It's an amazing restriction on the way the world works," Arkani-Hamed notes. All the particles observed prior to this are either spin-½ or spin-1. "After the Higgs was discovered," he says, "we added a new spin to that list: we finally found an elementary particle of spin 0."[50] Saying that this particle has zero spin means that it—unlike the other particles we're familiar with—has no special direction or orientation. It can be rotated in any way, and absolutely nothing will change.

Not only is the Higgs a new type of particle with a never-seen-before type of spin, but the interaction between a Higgs and other particles can be regarded as a new kind of force—one that takes its place alongside the electromagnetic, weak, and strong forces in the Standard Model, plus gravity, which presently stands apart from the others. Like the other forces, the "Higgs force" is mediated by a carrier particle, the Higgs boson, but is different from the other forces in that it is not based on gauge symmetry. Being new to physics as a verified phenomenon, the Higgs force is something that researchers are eager to investigate, although much higher-energy machines will be needed to probe it more fully.

"You often think that, once you've discovered something, it's an end," says Incandela. "What I've learned in science is that it's almost always a beginning."[51] Based on CMS's full 2012 data set, he says, "it is clear we are dealing with a Higgs boson though we still have a long way to go to know what kind of Higgs boson it is."[52] Is it the simple Higgs predicted by the Standard Model, or is it merely the lightest of several Higgs bosons predicted in other, follow-on theories, such as "supersymmetry" (to be taken up in the next chapter), which would extend well past the Standard Model?

And even if there is just one Higgs boson—a matter that is currently up in the air—physicists expect that there are many more particles to be found, although it will take high-powered devices, perhaps beyond those existing today, to uncover them. One chapter of particle physics, which has occupied researchers for more than a century, is drawing to a close—as the loose ends of the Standard Model are being tied up—just as an exciting new chapter is about to begin.

It seems inevitable—at least to Robbert Dijkgraaf, director of the Institute for Advanced Study in Princeton—that in the next couple of decades "there will be a place in the world where a new machine will go up, the largest 'microscope' ever assembled, which will enable us to peer further down into the structure of matter and closer than we've ever gotten to the Big Bang. The biggest unknown at the moment is about where that great adventure will take place."[53] And what machine of hitherto unmatched capabilities will become available to carry us into this new era of exploration?

Chapter 3

Beyond the Standard Model

THE HIGGS DISCOVERY WAS A SENSATIONAL TRIUMPH—the culmination of decades of progress in experiment and theory that provided a capstone to the Standard Model. The newfound boson can truly be called the "final piece" of a fantastic jigsaw puzzle to which physicists have devoted the better part of a century. That's because this "model" or theory is—like a hotel with no vacancies—fully occupied, having no room to accommodate even a single additional particle. "For the first time in our history," says Nima Arkani-Hamed, "we have a theoretical structure that describes all the interactions we know at accessible energies, which can be self-consistently extrapolated to exponentially higher energies."[1]

That statement, however, is not meant to imply that the work of particle physicists is done or anywhere near so. While "some might argue that there is no physics beyond the Standard Model," CERN researcher John Ellis points out, "history is littered with distinguished physicists (and others) who declared 'game over' prematurely." In 1894, for example, Albert Michelson claimed that "the more important fundamental laws and facts of physical science have all been discovered"—just prior to the discoveries of radioactivity and the electron. In 1900, Lord Kelvin similarly stated that "there is nothing new to be discovered in physics now; all that remains is more and more precise measurement." Shortly thereafter, Einstein proposed the existence of the photon and advanced the theory of special relativity.[2]

Particle physicists who are not seeking early retirement should take solace in the many formidable challenges that still await them. For starters, even though the Standard Model describes the behavior of all known particles to an unprecedented degree of accuracy—making predictions that have been confirmed many times over—its applicability is limited to a relatively small portion of the universe. One shortcoming of the model is that it does not include or have anything to say about gravity, the force that sculpts the cosmos and spurs the formation of structure—galaxies, galaxy clusters, and superclusters—on the largest scales. Furthermore, the laws of gravity, as laid out in Einstein's theory of general relativity, do not mesh well—and are, in some cases, incompatible—with the quantum theories that underlie the Standard Model. Physicists have long hoped to merge them into a single, unified theory of nature. While some progress has been made toward that end, it still remains a distant goal. And the Standard Model offers no path towards achieving this long-sought unification.

On top of that, astrophysicists now believe, based on multiple strands of evidence, that ordinary matter composed of quarks and leptons comprises less than five percent of the stuff that fills the universe. About twenty-seven percent, according to the latest estimates, consists of "dark matter"—particles created in great quantities during the Big Bang, which have properties different from those of the known particles. The Standard Model cannot explain this proposition, nor can it suggest any candidate particles that would make up this novel form of matter.

Instead, the motivation comes from astronomical observations and cosmology. Current theory holds that the inferred matter contributed to the formation of stars while providing the "gravitational glue" that holds galaxies and clusters together. It is called "dark" because we can't see it directly. Dark matter does not give off, reflect, or absorb light of any wavelength. We, therefore, have been unable to pin down the true nature of this shadowy substance although we can deduce its presence through astronomical observations. And those same observations tell us that there is simply not enough non-luminous, ordinary matter—such as dead stars, dust, or neutrinos—to keep structures like galaxies from flying apart. We need something new—something that's never been seen before—to handle the job.

On top of that, sixty-eight percent of the universe's mass and energy remains missing in action. It is thought to consist of equally mysterious dark energy that pervades all space, driving the accelerated expansion of the cosmos. When it comes to dark energy, physicists are quite literally in the dark. They don't know whether dark energy assumes the form of particles, and they haven't yet devised any accelerator experiments that are likely to shed light on its intrinsic properties. One of the few things we do know about dark energy is that it is not a part of the

Standard Model—a body of knowledge that has nothing to say on this particular subject.

Nor can the Standard Model explain the physics that gave rise to the Big Bang that ostensibly started it all. This owes, in large measure, to the fact that our theory of particle physics is presently divorced from our theory of gravity—general relativity. Furthermore, current teachings in cosmology tell us that the Big Bang created essentially equal amounts of matter and antimatter, but that the universe is now almost entirely matter—another critical circumstance that the Standard Model cannot account for and is, in fact, at odds with.

Even though this model arguably represents the most successful theory in the history of physics, the particles whose behavior it catalogues in vivid detail constitute less than five percent of the universe, leaving a lot of exploration yet to be done. In other words, says Joseph Incandela, "our work is 95 percent incomplete. We still have a long way to go to really understand the universe we live in." [3]

And while the Standard Model is supposed to be an essentially complete, self-contained theory, many basic questions can be asked about it for which the theory is ill-equipped to answer. "Its equations involve a score of numbers, like the mass of quarks, that have to be taken from experiment without our understanding why they are what they are," says Steven Weinberg. "If the Standard Model were the whole story, it would require neutrinos to have zero mass, while in fact their masses are merely very small, less than a millionth the mass of an electron." [4] How do neutrinos come by their diminutive masses? (While that question will be addressed in the next chapter, many physicists believe that its answer almost certainly involves physics beyond the Standard Model—or, at the very least, involves a modification of that model.) Why do the other elementary particles have the masses they do? And why are there three forces and three families of particles rather than, say, two or four?

The newly discovered Higgs boson, whose existence was long predicted, is in many ways the most enigmatic entity of all. This particle is seemingly connected with the electroweak phase transition—a cosmic change of state occurring in the early universe that resulted in formerly massless elementary particles acquiring mass. The Standard Model, however, cannot describe the nature of that phase transition, nor can it explain why there is such a huge gap between the electroweak energy scale and the Planck energy scale.

Many scientists believe that the Higgs boson can serve as a bridge to uncharted physics lying ahead, providing answers to some of the above questions while ultimately pointing the way toward a broader, more all-encompassing theory. With the Higgs discovery, noted Sergio Bertolucci, director of research and computing at CERN, "we have completed the known unknown. But we know it is not enough." And there is much more to be mined. [5]

The particle offers, for example, our principal means of understanding the Higgs field, about which our ignorance is profound, according to Harvard physicist Matt Strassler. "We do not know why [the field] is not zero, and we do not know why it interacts differently with different particles"—strongly with some particles like the top quark and weakly with others like the electron. Nor do we know whether there is, in actuality, just one such field or several. [6] Similar questions pertain to the Higgs particle, itself. Right now, its mass has been determined—with reasonable precision—by experiments at CERN. The best estimate of the Higgs mass as of March 2015, arrived at by combining data from the ATLAS and CMS experiments, is 125.09 GeV. This estimate has a margin of error of 0.24 GeV on either side, constituting a measurement precision of better than 0.2 percent, [7] which places it, according to CERN, "among the most precise measurements performed at the LHC to date." [8]

Despite the accuracy of that measurement, which is almost certain to increase in the future as the LHC accumulates more data, a deeper theory is still needed to calculate the Higgs mass from first principles. Furthermore, there are a number of questions about this particle we'd like to be able to answer. For instance, was the boson detected in 2012 the simplest possible Higgs, and just one of a kind, or is it part of a diverse ensemble, as some theories suggest? "Supersymmetry," an idea we will get to soon, predicts more than one Higgs boson—and five separate Higgs bosons in some models—including varieties that carry positive and negative electric charge. Other theories, of course, make different predictions.

The best way of ascertaining the Higgs's true identity is also the most obvious one—that being to examine it more closely, mapping out its properties as fully and accurately as possible. Physicists want to see how the particle behaves and exactly what it does. Towards this end, they need to—among other things—measure the new particle's mass and spin more precisely. They need to monitor its interactions and determine the rate at which it decays into other particles, comparing the numbers so acquired with theoretical predictions. "Only by learning the nature of the Higgs particle can we possibly understand the future focus of particle physics," says IHEP's Yifang Wang. [9]

"We've never seen a particle like this before," adds Arkani-Hamed. "It's like we've taken the first photograph of a totally bizarre creature. The picture is really blurry right now, and we need to sharpen the image." [10]

A major refurbishment of the LHC began in early 2013, with the accelerator resuming full operations in the latter part of 2015. More than 10,000 superconducting connectors that link the accelerator's magnets were reinforced, allowing the magnets to operate at higher field strengths that can produce more powerful proton collisions. The collider's voltage has also been increased and the cryogenics upgraded. By the time the revamped machine got up to full speed in June 2015, collision energies had nearly doubled, going from the 8 TeV achieved in its initial run to 13 TeV—a value close to its design energy.[11] And the LHC could achieve 14 TeV collisions before the end of "Run 2," as this second phase is being called.[12] The number of proton collisions will increase as well, boosting the amount of data to be collected and thereby making the LHC ten about times more sensitive—and ten times better able to observe rare phenomena.[13] Physicists, meanwhile, intend to acquire more precise measurements of the Higgs boson—an objective that is virtually guaranteed.

There is, of course, a further expectation for the upgraded facility. Physicists hope that some new particles will be seen by the LHC. But even if that happens, says former CERN head Luciano Maiani, "it is very unlikely that the LHC will see everything. It's more likely that the LHC will be like a lamp in the darkness that is able to illuminate the tail of the animal that we believe is there but without completely identifying its nature."[14]

Arkani-Hamed agrees that we are unlikely to see "the complete story" at the LHC in going from 8 to 13 TeV. In light of that, the prudent strategy is to start making plans for much higher precision and much higher energy machines, like the 100 TeV collider being planned in China.[15] A facility on that scale would not only increase our familiarity with the Higgs boson and the Higgs field that spawns it, but would also have a much greater chance of uncovering new particles that lie outside the Standard Model, serving as a gateway to unexplored avenues in physics.

China's proposed collider, which will be discussed at greater length in Chapter 5, is designed to pursue both of these broad objectives. The machine would operate in two stages, first as a "Circular Electron-Positron Collider" (CEPC) running at collision energies of about 240 GeV. It would later be converted into a "Super Proton-Proton Collider" (SPPC) that would make use of the same tunnel and provide much higher collision energies (owing to the proton's greater mass), possibly as high as 100 TeV.

While the proton collider holds the potential to discover new particles, more massive than any ever seen before, a lower-energy electron-positron machine would offer its own set of advantages—advantages that would also apply to another proposed machine, the more than 30 kilometer long International Linear Collider (ILC), which Japan is interested in hosting. (The CEPC, SPPC, and ILC will also be discussed in Chapter 5.)

The main benefit for experimentalists is that collisions of electrons and positrons are much cleaner than collisions of protons. Almost everything physicists have learned so far is consistent with the notion of electrons and positrons being simple, point-like particles that have no inner structure. Protons, by contrast, are more complex and chaotic objects, sometimes described as bags of quarks and gluons, with additional, virtual particles constantly popping into and out of existence, borrowing energy and returning it as they come and go. That makes precise measurements much more difficult in the wake of proton collisions because all those extra parts strewn about can obscure the picture and mask subtle effects.

Oxford physicist Brian Foster, the European regional director of the ILC project, compares proton collisions to smashing two oranges together at high speeds. Suppose you're really interested in seeing what happens when the seeds from two different oranges meet head-on, but that may be hard to discern with all the juice, pulp, and other orange bits scattered throughout the collision zone. [16]

Electron-positron colliders of sufficient energy, operating well above the 125 GeV mass of the Higgs boson, can be finely tuned to churn out these particles in great quantities, which is why such machines are sometimes referred to as "Higgs factories." Moreover, they can be used to perform measurements of a precision that is simply not attainable at a proton collider like the LHC (or the proposed SPPC). Columbia physicist Brian Greene likens the LHC to "a sledgehammer, slamming protons together to create whatever new stuff we haven't seen before. It is really good at that." An electron-positron collider, on the other hand, is more like "a scalpel, a precision instrument capable of much more fine-grained analysis." It could be used to pin down specific particles and energies that deserve closer scrutiny, he says, and "that is the machine that we want to start building now." [17]

Yifang Wang agrees, but he'd especially like to see a machine of that general sort (although circular rather than linear in design) take shape in China. An electron-positron machine can provide us with a lot of precision measurements that are desperately needed, Wang says. "The Higgs is the most important particle in the Standard Model, and we know much more about all the other particles in that model. We have, for example, very precise measurements of the electron, and we need measurements of the same precision for the Higgs." [18]

As stated before, there are many mysteries surrounding the Higgs boson that physicists hope to clear up, perhaps none as baffling as that pertaining to the particle's mass, which LHC measurements recently placed at just over 125 GeV, give or take. One problem that might keep theorists awake at night is explaining why the Higgs mass is so small when calculations based on quantum field theory—the theory upon which the Standard Model rests—suggest that the Higgs boson should be immensely heavier than that. Those calculations, in fact, would seemingly push the Higgs mass upwards, by a factor of 10^{16} or 10^{17}, towards the so-called Planck energy and mass scale of about 10^{19} GeV. The Planck scale, which corresponds to tiny lengths on the order of 10^{-35} meters, is pretty much the end of the line so far as quantum field theory and general relativity is concerned—the point that we, seemingly, cannot go past in terms of increasing energy or decreasing distance. To proceed further in those directions, new physical laws will be needed to guide us. The Planck scale represents the energy and distance thresholds at which gravity breaks down as a continuous force. It is also the scale at which gravity becomes comparable in strength to the other forces and its quantum effects must be taken into account. In other words, it's a transitional point where a quantum theory of gravity needs to take over, just like the quantum theories that currently describe the electromagnetic, weak, and strong forces.

But why do theoretical arguments suggest a far higher Higgs mass—one that is much closer to the Planck scale than what was observed at CERN? The answer lies in peculiar quantum mechanical effects that are not especially intuitive to non-physicists. A subatomic particle like the Higgs doesn't sit alone doing nothing. It flies around at high speeds and, because it has mass, it will interact with the Higgs field like other massive particles. But it will also interact with other particles that it encounters in its journey through space and sometimes will even turn into other particles, briefly, before reverting back to a Higgs. It might, for example, temporarily turn into a top quark-antiquark pair or into a pair of W bosons or even into a pair of extremely heavy virtual particles before resuming its life as a Higgs. Each of these interactions, conversions, and identity switches would make a contribution—a quantum mechanical contribution—to the Higgs mass that should, in the end, bring it close to the realm of 10^{19} GeV.

Of course, this was not seen at the LHC. Nor would it have been possible to detect a particle anywhere near that heavy at a collider rated at only about 10 TeV (or 10^4 GeV). This is one conundrum that researchers are currently grappling with—how to make sense of the particle's low mass (even though, ironically, the mass that was measured fell squarely within the range of predicted values). It appears that somehow the quantum contributions from the various particles are cancelling each other out instead of adding up, although how that happens is almost a total enigma. Passing this off as a coincidence is not a satisfactory explanation to most physicists who see that, instead, as an example of "fine-tuning" in which someone, or some unseen force, appears to be adjusting nature's dials to an almost absurd level of precision.

"The required fine-tuning, which must have sixteen-digit [or sixteen decimal point] accuracy, is more extreme than the fine-tuning required to make your pencil stand on end," Lisa Randall says. And someone or something must be using hidden wires, so thin as to be invisible to us, to keep that pencil standing up in defiance of common sense, gravity, and the laws of probability. "Fine-tuning is almost certainly a badge of shame reflecting our ignorance." [19] Particle physicists, accordingly, would prefer to find a more natural and logical explanation for why our universe is the way it is.

Another way to cast this dilemma is not in terms of the Higgs boson mass *per se*, but rather in terms of the value, or energy, of the Higgs field. The continual appearance and disappearance of virtual particles will make quantum contributions to the Higgs field, increasing its strength or energy towards the Planck scale, in the same way that the winking in and out of virtual particles contributes mightily to the proton mass (as discussed in Chapter 2). Since elementary particles—including the W, Z, and Higgs bosons—derive their mass from interactions with the Higgs field, having a Higgs field up near the Planck energy would inevitably lead to the presence of superheavy elementary particles of all sorts—particles so massive they would be at risk of collapsing into black holes. In other words, the electroweak scale of about 250 GeV—which is also, not coincidentally, the value of the Higgs field in empty space to which the mass of Standard Model particles is tied—would be shown to be much closer to the Planck scale, not separated by a gap of more than sixteen orders of magnitude. This large disparity, which strikes many physicists as odd and "unnatural," is sometimes known as the "hierarchy problem."

The hierarchy problem can be cast in various ways. It ties into the surprisingly low strength of the Higgs field and low mass of the Higgs boson—and of the other known elementary particles—as compared to the Planck energy and mass scale. It can also be framed in terms of the discrepancy in strength between gravity and the other forces—electromagnetic, weak, and strong. In fact, the gravitational force between two electrons is about 10^{43} times weaker than the electromagnetic attraction they experience. [20] These forces would match if electrons were more than ten quadrillion times heavier—close to the Planck scale in mass, in other words. [21]

But why are the forces of nature so different in strength, and why is the chasm—separating known particle masses from the Planck scale mass—so vast? This, again, leads us back to that same hierarchy problem that's been gnawing away at particle physicists for decades. And many of them are confident that solving this problem could offer a path toward uncovering new physics and new associated particles that lie beyond the Standard Model. [22]

The thrust of this inquiry can be centered around the Higgs boson itself. If the huge quantum mechanical effects, which are expected to beef up the Higgs mass, somehow offset each other, just how does this "offsetting" come about? Getting to the bottom of this quandary is not an optional exercise—it's mandatory, argues Fermilab physicist Don Lincoln. If the particle discovered in 2012 is, in actuality, the Higgs boson and its mass is 125 GeV, as experiments have shown it to be, says Lincoln, "this is completely unnatural in the Standard Model, and we *must* discover some new physics to explain it." [23]

Fortunately, an answer, or partial answer, may reside in the idea that the universe has an additional symmetry, called supersymmetry, that hadn't been recognized before. Rather than being a novel theory of the universe, supersymmetry is better described as a feature of some theories of the universe—a proposed new principle that, if proven correct, might someday stand alongside other hallowed principles like the conservation of energy and momentum. Significant development of supersymmetry began in the early 1970s. Some regarded it as a rather abstract mathematical notion, which revealed, as Edward Witten puts it, "a new dimension of space and time that could only be described in quantum mechanical terms." [24]

But supersymmetry also has more tangible consequences for physics. By 1981, the concept had evolved—at least in one version proposed by the physicists Savas Dimopoulos of Stanford and Howard Georgi of Harvard—into an extension of the Standard Model, complete with predictions that might someday be put to an empirical test. If nature is ultimately shown to be supersymmetric, says University of Michigan physicist Gordon Kane, "the Standard Model will not be wrong but will simply become part of a more complete description of nature. That is the way science progresses." [25]

One important part of this "more complete description" is the notion that every particle in the Standard Model has a heavier and as-of-yet unseen counterpart—a proposition that would essentially double the number of elementary particles in the universe. More specifically, supersymmetry holds that every boson in the Standard Model (such as a photon, gluon, or Higgs particle) has a fermionic "superpartner" whose name—by an agreed upon convention—ends in "ino" (such as photino, gluino, or Higgsino). Conversely, every Standard Model fermion (such as a quark, lepton, electron, or neutrino) has a bosonic superpartner whose name—again by convention—begins with an "s" (such as squark, slepton, selectron, or sneutrino). The superpartners would have the same mass and charge as their known companions, but their spin would differ by a half integer. (That last condition makes sense given that bosons have whole number spin whereas fermions have half-integer spin.)

None of the known bosons and fermions has the right characteristics to qualify as the superpartners posited by supersymmetry, which is another way of saying that none of these presumed particles have been discovered yet. If they really exist, experimental physicists should have plenty of work ahead of them, with roughly a dozen and a half new particles to find, study, and characterize. The six quarks, six leptons, and five bosons (photon, gluon, and Z, W, and Higgs) of the Standard Model would be matched by at least 17 additional superpartners and possibly more because a normal particle can, in principle, have more than one superpartner.

Supersymmetry describes an intimate and hitherto unsuspected link between force-carrying particles (or bosons) and matter particles (or fermions), thus representing a new symmetry of nature that unifies force and matter. The theorists who conceived of this idea showed, to the surprise of many, that the equations of physics should work in the same way, even when the terms for the bosons and fermions that appear in those equations are interchanged.

As an analogy, one might consider twenty-two soccer-playing girls consisting of eleven pairs of identical twins. For the sake of argument, let's make the somewhat unrealistic assumption that a girl and her twin truly are *identical*, having the same size, competitive drive, playing ability, and so forth. Let's further assume that the girls are divided into two teams, dubbed the Bosons and the Fermions, with each pair of twins separated from each other. Play begins, and the teams—as might be expected—are evenly matched. If, in the middle of the game, the coaches start swapping players—exchanging a girl from one team with her twin sister on the other team—the outcome of the game should not be affected in a discernible way. Everything would unfold exactly the same. That is similar to what theoretical work on supersymmetry has shown: Swapping bosons for fermions with otherwise identical properties should not change the outcome of a given physical encounter.

However, we already know, from abundant experimental evidence, that one aspect of supersymmetry cannot be true—namely that the masses of a particle and its superpartner cannot be the same: The superpartner must be heavier. If, for example, the selectron had the same mass as the electron, and the up squark had the same mass as the up quark, both would have been discovered long ago. Furthermore, if selectrons had the same mass as electrons, they would have infiltrated atoms, altering chemistry and interfering with the formation of structure in a radical way—to the point that the world as we know it would, on the contrary, be a world unknown. The possibility of selectrons having a mass equal to that of electrons is, therefore, positively and absolutely ruled out, implying that if supersymmetric particles exist, they must be heavier—and perhaps substantially so—than their known counterparts.

Another way of putting it is that if supersymmetry is realized in nature, it must be a "broken" symmetry. The difference in mass between a Standard Model particle and its superpartner provides an indication of how badly supersymmetry is broken.

After being broken, a symmetry doesn't disappear altogether. Instead, it becomes difficult to recognize but is still there, lurking behind the scenes in a diminished form. Physicists are not put off by the notion of broken symmetry, as the current consensus holds that electroweak symmetry is broken too. Indeed, examples of broken symmetry appear to be common in the world around us—and in the world we stand on. The Earth, to take a nearby example, enjoys some of the attributes of spherical symmetry, but that symmetry is imperfect because our planet is not a perfect sphere. The human face tends to be fairly symmetric when comparing the left and right sides but never exactly so. Upon close scrutiny, some differences in shape and features—such as freckles, blemishes, wrinkles, or crease lines—can always be seen. [26]

Another example of broken symmetry would be a child's spinning top with a minor design flaw, which makes it slightly heavier on one side than another. This defect will make the top wobble sooner than it would have otherwise and also more likely to end up falling onto one side in particular, that being the heavier side. The top still exhibits some degree of rotational symmetry as it spins about its central axis but not as much as a flawless, completely uniform top. [27]

So why has supersymmetry captured the attention of thousands of physicists over the decades given the restrictions that must be imposed on it—the fact that even if this phenomenon is real, we know from the outset that it's already broken and otherwise flawed? One reason is that supersymmetry offers the potential for alleviating both the hierarchy problem and the mystery of dark matter, despite the fact that it was not originally developed for the purposes of addressing either of those issues or any other theoretical inconsistency in physics. Supersymmetry models predicted that the Higgs mass should be below 130 GeV, as turned out to be the case, [28] and predicted a decade before its discovery that the top quark should be significantly heavier than other quarks and leptons.

But supersymmetry is much broader than that, offering a possible rationale for the masses assigned to different particles and for the strengths of the different forces, simultaneously explaining many of the properties of our universe. [29] Physicists over the decades have been enamored with this idea because of its expansive scope and the deep, beguiling mathematics from which it springs. Incorporating supersymmetry into the fabric of physics makes it easier for practitioners to solve a wide variety of problems. Moreover, the fact that supersymmetry solved problems—or offered solutions to problems—"that it was not introduced to solve was itself a powerful hint that it was indeed part of the description of nature," Gordon Kane has argued. [30]

Let's first consider what supersymmetry might do for us regarding the hierarchy problem—an issue that has long weighed (and preyed) on the minds of physicists. The basic premise is that the new (supersymmetric) particles would drive down the mass of the Higgs boson and the energy of the Higgs field almost exactly as much as the known (non-supersymmetric) particles would push them up. The way that works is astonishingly simple. For starters, we'll repeat one feature of supersymmetry—namely that every Standard Model boson is paired with a fermionic superpartner just as every Standard Model fermion is paired with a bosonic superpartner. Bosons boost the energy of the Higgs field when they interact with it, which also makes the Higgs particle (and other elementary particles) heavier. Fermions, on the other hand, make a negative contribution to the Higgs field energy and, by extension, to the Higgs field mass. The superpartner fermions and bosons, in other words, will counteract the effects of the Standard Model fermions and bosons. The cancellation is not quite perfect, however, because the superpartners are heavier than their counterparts owing to the aforementioned broken symmetry. The Higgs mass and Higgs field energy can be held in check, and kept from running amok, so long as the mass of the superpartners is not too big.

"What supersymmetry does for you is that it stabilizes the mass of the Higgs boson and keeps the quantum corrections under control, but you still need to get the right value [of the Higgs mass] in the first place," Ellis says. "If you start off with a small value for the Higgs mass, it helps you stay small though it doesn't explain how you got the small value in the first place."[31]

A possible solution to the hierarchy problem would be big news in itself, but supersymmetry can potentially do more—much more. For if our universe were indeed supersymmetric, superpartners as well as the familiar particles of the Standard Model would have been produced in great numbers during the Big Bang. Being highly unstable, the superpartners would quickly decay. A stipulation of many supersymmetry models, which was designed to make the theory consistent with observations, holds that supersymmetric particles are always created in pairs and always destroyed in pairs—never created or destroyed in isolation. This proposition can be restated in a slightly simpler way: when a supersymmetric particle decays, another supersymmetric particle is always made. You can't start with one superpartner and, after various processes and transactions, end up with none. Furthermore, a superpartner—and any other particle, for that matter—can only decay to a lighter particle (or else some longstanding and cherished conservation laws would be violated).

The logical conclusion to be drawn from this is that superpartners will keep decaying until, at some stage in their decay chain, they arrive at the final step whereby the lightest possible superpartner is produced. All told, large quantities of this kind of particle would be produced and spread throughout the universe. But once the decay products have reached the lowest attainable mass, there can be no further decays because that would leave zero superpartners—in violation of the principle stated above. So the lightest superpartner must be completely stable. And, because we haven't yet seen such a particle in accelerator experiments or in astrophysical studies, the presumption is that the lightest superpartner must also be weakly interacting and, therefore, difficult to spot. To stay concealed as long as it has, this particle must not have any electric charge, nor would it be susceptible to either the electromagnetic or strong forces—features that would otherwise have made it readily detectable.

While a particle of this sort would not participate in the formation of stars and galaxies, it would be a prime candidate for the dark matter that is thought to populate the universe. At the moment, the dark matter candidate most favored by theorists is the "neutralino"—a class of four particles that consist of varying mixtures of the superpartners of the photon ("photino"), the Z boson ("zino"), and the Higgs boson ("Higgsino").

This explains part of supersymmetry's appeal and one reason why it has captivated physicists for decades. For if nature actually abides by this symmetry, it would give us twice the number of elementary particles while offering a way out of the hierarchy and dark-matter conundrums. Perhaps some of the universe's more puzzling aspects stem from the fact that we've only been looking at half of the picture. "Supersymmetry provides a single solution to all of these problems in a way that is undeniably simpler, more elegant, and more beautiful than any other theory to have been proposed," claims Fermilab physicist Dan Hooper. "If our world is supersymmetric, all of the puzzle's pieces fit together nicely. The more that we study supersymmetry, the more compelling the theory becomes." [32]

It is an enticing proposition, to be sure. However, until we get some experimental proof, we can't know whether supersymmetry, beautiful as it may be, is actually written into the laws of nature or not. Up through the LHC's first run, which came to a close on February 14, 2013, no solid evidence has been mustered in support of this wondrous idea, and a fair number of physicists are now expressing their doubts. But few researchers are ready to abandon the hunt just yet.

There are numerous strategies investigators can pursue in their search for superpartners—the lightest of which, as indicated before, might comprise the elusive dark matter. One approach, which is being pursued at the LHC, is to produce Higgs bosons in large quantities and look for rare, unexpected decays that might lead to the discovery of unknown particles, perhaps including light supersymmetric particles. "The light supersymmetry sector will still be hard to explore at the LHC," maintains Stanford theorist Michael Peskin, who believes that electron-positron machines can do a better job of searching for exotic Higgs decays than a hadron collider. [33] He also believes that electron-positron machines are superior for directly making dark matter particles, unless those particles happen to be so massive that a hadron collider is required for their production. [34]

Harvard's Matt Strassler agrees. Given its higher precision, an electron-positron machine, such as the proposed ILC or China's CEPC, could do an even better job of observing rare interactions and—in the process—finding hints of never-before-seen particles. The CEPC could be a real "discovery machine," not just a Higgs factory, he says. "So this kind of machine has a niche, bringing us to the question: What do we do with the Higgs now that we've found it? Well, you want to check precisely what the Higgs is supposed to do, but you also want to check what the Higgs is *not* supposed to do. Here you could pick out things that occur rarely, say one time out of a million. When you find them, you might be discovering new, light particles, and these particles would, by definition, lie outside the Standard Model." These particles, which may not be produced by any other means, could provide the keys to unlocking deep secrets about nature, such as what dark matter really is. And the only way to discover them may be through the decay of the Higgs, which could be difficult to observe at the LHC because of all the background noise. [35]

Proton-proton colliders, owing to the higher energies they can attain, would aid the search for heavier superpartners that are beyond the reach of an electron-positron machine. Investigators must cast a wide net because supersymmetry predicts the existence of a host of new particles without specifying the energy levels at which these particles can be found. And supersymmetry, again, is not just one theory; it can take many forms and manifest itself in many different ways. A large number of models, accordingly, have been put forward. But there is a common thread to most of the search strategies: After smashing protons together, physicists rely on detectors to help them keep track of all the energy of the known particles emerging from a collision. If any superpartners are produced, they will decay almost immediately. The only trace of their presence will be "missing energy" and "missing momentum"—an imbalance between the energy and momentum going into the initial collision and that seen coming out. These two quantities—the input to a collision and the output from it—should, in principle, be equal. Any disparity between the two would be a sign that observers are missing part of the story.

When the quarks, antiquarks, and gluons inside protons collide, they create a cascade of particles that might conceivably include the superpartners of the original players: squarks, antisquarks, and gluinos. These superpartners, in turn, would eventually decay down to the lightest possible superpartner—perhaps neutralinos, being one of the most popular dark matter candidates at the moment. Researchers are looking for a particular signature, a high-energy jet of quarks or other known particles shooting off in one direction—say to the left—with nothing coming out to the right. Such an imbalance, or hint of missing momentum, could offer us the first direct evidence of supersymmetry—and maybe, by extension, of dark matter as well. "A so-called monojet event like this would look odd," comments Masahiro Morii, chair of the Harvard Physics Department and a member of the ATLAS team. "It would appear to violate momentum conservation, and what it's telling us is that something invisible went the other way. When we talk about dark matter searches at the LHC, this is almost always the most powerful probe—looking for monojet signatures, be it mono-photons, mono-Z bosons, mono-Higgs bosons, and so forth." [36]

"You know the little saying, 'What is the sound of one hand clapping?'" Ellis asks. "Well, what is the sound of one jet of particles? It could be dark matter." [37]

In preparing for the LHC's Run 2, which started up with some relatively low-energy proton collisions in May 2015, Morii explains that experimentalists group supersymmetric particles by how easy it is to make them. "We know that quarks and gluons have strong interactions, and if I apply supersymmetry to that, we get squarks and gluinos." This, he says, is where the initial effort will go. The search for squarks and gluinos will be the focus of the first year or so of operations after the restart. "If we find either of them, that will be great. If we don't, we'll go after the less easy stuff."

High on the agenda will be the search for the lightest supersymmetric particle or, equivalently, the lightest superpartner, both of which share the same acronym, LSP. Most models suggest that the LSP is presumably a neutralino, which can be made from quark-antiquark collisions—as well as from the decay of a top quark's superpartner, which is called the top squark (or "stop") for short. "We don't know how massive the top squark is, but we think it has to weigh at least hundreds of GeV," Morii says. "The nice thing about supersymmetry is that even if we don't know the masses of the particles we're looking for"—and really have no way of calculating them in advance—"we still know their couplings because the couplings of the top quark and top squark with the Higgs field should be identical."

The "couplings" to which Morii refers relate to how strongly a particle interacts with the Higgs field. The strength of a charged particle's interaction with the electromagnetic field, by way of example, depends on the particle's electric charge. A particle's experience of the strong force, similarly, depends on the amount of "color" charge it carries. Physicists also believe that the degree to which a particle—such as a quark or lepton—interacts with the Higgs field depends on a property of that particle that is analogous to charge and color, although they don't yet know exactly what that property is. They further assume that the strength of this interaction will be the same for a particle and its superpartner, which is predicated on the conviction that this aspect of supersymmetry is not broken. Drawing on what they have learned about ordinary (Standard Model) particles, physicists can then determine the probability of making various superpartners and also predict how these supersymmetric particles should decay. "In this respect," Morii says, "the theory has relatively little wiggle room." [38]

But they still have to look to make sure that the clues they've divined from theory are pointing in the right direction. No one can be certain, of course, as to what will turn up now that the LHC has reopened and started firing at 13 TeV. "If we knew what we were going to find, we'd all just go home," says Harvard physicist Melissa Franklin who's also on the ATLAS team. [39]

The suspense is considerable, as physicists wonder what the next round of data taking will reveal. "There are no bad answers here," opines Arkani-Hamed. "We just don't know which way nature is going to go." [40] If we continue to see the Higgs and no other new particles, that would put current ideas under strain—and possibly to the breaking point—which is not necessarily bad. When the Michelson-Morley experiments of the 1880s failed to see the aether—the long-postulated medium for the propagation of light—Arkani-Hamed says, "that heralded a paradigm shift in physics." [41]

Of course, he would be thrilled if something spectacular happened such as the detection of a bona fide superpartner. That would not only provide a stunning confirmation of a decades-old idea but it would also ensure that there are a large number of new particles to be found, perhaps close at hand. Any debate over whether to build a bigger, brawnier collider would end instantly and about the only questions remaining would be how fast can you build it. "Discovering supersymmetry would be one of the most amazing things we've seen in physics since the early part of the 20th century," Arkani-Hamed adds. [42] And it would surely point to a bright and exciting future for particle hunters and theorists alike.

Such a discovery, adds his IAS colleague Edward Witten, "would be one of the real milestones in physics, made even more exciting by [supersymmetry's] close links to still more ambitious theoretical ideas. Indeed, supersymmetry is one of the basic requirements of string theory, which is the framework in which theoretical physicists have had some success in unifying gravity with the rest of the elementary particle forces." [43] In fact, it has been proven mathematically that supersymmetry is the only symmetry that can be added to general relativity without making Einstein's theory incompatible with the world we inhabit. It may, in other words, be the final symmetry of space and time. [44] Moreover, the only way that string theory can reproduce the particles of the Standard Model, so far as we presently know, is by incorporating supersymmetry. [45] Therefore, Witten adds, "the discovery of supersymmetry would surely give string theory an enormous boost." [46]

But if nothing turns up during LHC's second, three-year run, what might that mean? It could lend more weight to the argument that our universe is not supersymmetric and that the quantum cancellations that keep the Higgs mass from going haywire are the result of fine-tuning—pure chance—rather than an additional symmetry of space and time. Another possibility is that the LHC, even operating at the elevated realm of 13 or 14 TeV, is simply not powerful enough to produce these superpartners and dark matter particles whose masses, as noted, are still unknown. Perhaps they are in the vicinity of hundreds of GeV or one thousand times heavier. If the latter case prevails, we will need a higher-energy collider, perhaps on the order of 100 TeV, like the collider that is currently under review in China or the collider of a roughly comparable scale that is being considered by CERN. (More will be said about both of these machines in Chapter 5.)

"There's a good chance that they [dark matter particles] would be produced at the LHC but an even better chance they would be produced at a 100 TEV collider," says Ellis. [47] He believes that patience and a long-term perspective are in order, given that there are many possible versions of supersymmetry, which allow for a wide range of superpartner masses. It took 48 years to discover the Higgs boson after that particle was proposed, he says, and it might take a similar amount of time, or longer, to confirm supersymmetry—an idea that dates back at least to 1971.[49]

"You don't always find things on the first look," Fermilab physicist Joseph Lykken said in 2013, after the LHC had concluded its first round of data taking. "Sometimes you find them on the last look. We didn't have a theory that said supersymmetry had to look exactly like this; we had a range of millions of different models to look at. So I don't want to be too discouraged."[50]

But by that time, some physicists were already concluding that supersymmetry alone might not be able to solve the hierarchy problem. Based on what the LHC did see—and, more importantly, what it did not see—by February 2013, many felt that the superpartners (if there were any to be found) might be too massive to bring about enough cancellations to hold down the Higgs mass. Although supersymmetry could still make important contributions on this front, it might have to be combined with other physical effects. An additional ingredient might be required for supersymmetry theories to succeed, Randall says, and that "extra ingredient might be extra dimensions."[51]

By now, many non-physicists are used to the idea that there are three spatial dimensions plus a dimension of time, which all come together to form a four-dimensional space-time. But what if our universe had hidden "extra" dimensions, in addition to the four dimensions mentioned above, that we cannot see either because they are too small or because our view of them is obscured for other reasons?

There are some familiar parallels to this in everyday life. Imagine, for example, a tightrope extended between two trees. To a person trying to get from one tree to the other, the tightrope is a one-dimensional object, and his or her motion is constrained to that single dimension—either moving forward towards the second tree or backward towards the starting point. That's the way it would appear to most people in that situation, trying to make their way across: They probably would not be thinking about how an ant, for instance, with other options available to it, might view the situation. The ant could, like the tightrope walker, make its way along the rope, moving directly from one tree to the next or back. But, with its small size, it could also move in different directions while on the rope—perhaps making small loops around a particular spot on the rope, or spirals around the rope, as it moves forward or backwards. To the ant, this rope is a two-dimensional object because it, unlike us, has the ability to move in two independent directions.

Such is an illustration of how a hidden or extra dimension might exist in nature, of which we humans aren't aware simply because we're big and that dimension is very small. And there may not be just one extra dimension but many. String theory (and its relative, M theory), for instance, posits that there are six (or seven) extra spatial dimensions that are so small, and curled up so tight, that we cannot see them nor could we ever wander around inside. But you don't have to believe in string theory to accept the possibility that our universe might have one extra dimension or more. This idea could help solve the hierarchy problem in a manner that's different from what supersymmetry has to offer. That problem, once again, relates to the huge gap—of some sixteen orders of magnitude—that separates the electroweak scale from the Planck scale. Although the hierarchy problem can be discussed in terms of the bafflingly low Higgs mass, it can also be expressed in terms of the apparent weakness of gravity—and the fact that the gravitational force between two electrons (as touched upon previously) is some 10^{-43} times weaker than the electromagnetic force between them. But what if gravity spreads its influence not only in the four dimensions familiar to us but also in the extra-dimensional realm? It might appear feeble to us, stuck in the four-dimensional world, simply because we can't see its full effect. The more dimensions there are, the more diluted gravity would get and the weaker it would look to us. If we found out that there were extra dimensions in which gravity operated too, we might also recognize that the gap between the Planck scale and the electroweak scale was not quite as big as we thought. The low Higgs mass would no longer be quite so troublesome to theorists, nor would the hierarchy problem seem so problematic after all.

But if these extra dimensions have been hiding from us all the while, how might we finally see traces of them and how, specifically, might a particle accelerator like the LHC, or a more powerful successor such as China's proposed collider, help the cause? It's possible that machines of this sort could turn up exotic heavy particles associated with extra dimensions called Kaluza-Klein (or KK) particles. These hypothetical particles are named after the German mathematician Theodor Kaluza and the Swedish physicist Oskar Klein who, as of 1919, began developing a unified theory of gravity and electromagnetism that needed an extra dimension in order to fit the forces together. Without this additional dimension, there simply was not room in the governing equations to accommodate a combined theory.

One way to picture how such particles might arise is to think of a barge traveling in a uniformly narrow and shallow canal that runs from Albany to Buffalo. [52] This barge is so big compared to the waterway that it almost completely fills the canal it passes through. It's such a tight squeeze, in fact, that the barge cannot move sideways or up and down at all. Like the tightrope walker from our previous example, its motion is confined to a one-dimensional path—proceeding east towards Albany or west towards Buffalo.

Let's now imagine a boat that's slightly narrower. If you saw it motoring along to Buffalo, you might not notice that it is actually moving in two dimensions, with small though rapid and steady jostles from side to side, as it heads west. Because of those lateral movements, the boat has additional kinetic energy.

The boat also does not extend as far beneath the surface as the barge, so that in addition to moving forward and backwards, and from side to side, it also bobs up and down. This narrower, shallower boat is thus moving in three dimensions, not just one, and there is kinetic energy associated with its vertical motion too.

Instead of barges or boats navigating through a canal, let's now think of a particle traveling at close to the speed of light through an underground tunnel— a circular track of 27 kilometers, or maybe even 100 kilometers, in diameter. The people who built this huge racecourse might have assumed that particles accelerating around it were restricted to that single loop—a sizeable one, to be sure, but a one-dimensional circuit all the same. What if, while speeding around that giant circle, a particle created during a collision was also moving in an unseen dimension, making tiny, ultrafast loops in a direction perpendicular to its prescribed route of travel. Such a particle would appear unexpectedly heavy owing to the added energy of motion it acquired during its constant gyrations through the hidden dimension. Particles like it might even make more rapid, higher frequency loops through this concealed realm, speeding up in regular ("quantized") multiples. They would carry even more momentum and, to us, would appear heavier still.

No experiments to date have turned up any signs of a set of strange particles that fit this pattern. However, if that were to happen, and if physicists at the LHC or a next-generation collider managed to detect a slew of exotic particles like this, similar to each other while exhibiting a regular pattern of increasing masses, they might also have spotted the first clues of an extra dimension (or two)—a finding that would revolutionize physics in addition to offering some relief for the hierarchy problem. (And science fiction writers, who have faithfully believed in extra dimensions, might find some vindication too.)

Until one of the ideas put forth unambiguously fits the data, neatly tying up all loose ends, theorists have a job to do. And that's to keep turning out more ideas until one of them sticks. Yet another way around the vexing Higgs mass question would be to suggest that the Higgs is a composite particle—and thus has a substructure—in the same way that hadrons (like protons and neutrons) are composite. If this speculation proves correct, and so far there is not a shred of evidence to support it, the Higgs boson would not be an elementary particle. It would get its mass in the same way that a proton gets its mass—from the combined masses of its constituent parts, which are constantly moving around, and from the energy that binds them together. A multifaceted Higgs boson of this sort would not be subject to the same quantum effects that would drive up the mass of a fundamental Higgs particle. The appearance of a lightweight Higgs, in this scenario, is nothing to get worked up about. And the whole crisis would thus be averted.

One way to explore the possibility of a composite Higgs would be to measure the processes that result in making pairs of Higgs bosons. "The LHC could not do that," Strassler says, "because the production rate would be too small"—and the phenomena too rare—"but a 100 TeV machine could." That would afford the opportunity to see how the Higgs interacts with itself, which, in turn, would provide some indications as to whether the Higgs boson seen in 2012 is the simplest possible Higgs or if it might have a concealed substructure. [53]

Three Higgs particles could, for example, interact with each other, which is something that has never been seen before, notes Arkani-Hamed. One way this might occur is for two Higgs to collide into each other to produce another Higgs—most likely a heavier "virtual" one—which would, in turn, quickly decay into two other particles (such as W or Z bosons). Studying a "triple Higgs" interaction of this sort is crucial to understanding whether or not the Higgs is point-like, having no underlying structure that we can resolve. "This is an absolutely critical measurement that only a 100 TeV machine could do," Arkani-Hamed says. [54]

Current theory hinges on the premise that the Higgs particle is point-like. But evidence to the contrary, suggesting that it is made up of smaller pieces, would rock the physics world. "It would really open up the onion," as Arkani-Hamed puts it. [55] Not only would it change our view of the recently spotted Higgs boson, it would also raise the prospect that other supposedly fundamental building blocks of nature might have smaller constituents as well. New theories would be required to account for this phenomenon, and particle physics would never be the same again. And that, to paraphrase Kane, is how science progresses.

As can be seen, there are a number of ways that the puzzle concerning the Higgs mass, which launched us on this line of inquiry, might be set right. "It could be supersymmetry or extra dimensions or some other model, such as a composite Higgs," Yifang Wang says. "We need one of these ideas to pan out, or some combination of them, or maybe something that's not yet been invented. We don't know which option, if any, is correct. That's why we need a new accelerator to tell us which way to go." [56] And Wang, of course, has some ideas as to the optimal location for such a facility.

Naturally it's going to take a huge investment to construct a facility of that scale, which would likely cost upwards of $10 billion. Justifying an outlay like that won't be easy since there are no guaranteed discoveries in a venture like this—especially when the questions at stake, regarding dark matter and the origins of mass scales in physics, are so big. No one at this juncture can confidently say: Build this device and we are certain to find x, y, and z. About the only guarantee we have is that if we want to want to comprehend our universe at its most basic level, we have to keep looking—we have to continue to examine nature at higher energies and at smaller distance scales.

Nima Arkani-Hamed of the Institute for Advanced Study is a theorist and phenomenologist who has given considerable thought to the exploration of physics beyond the Standard Model. (Photograph by Andrea Kane, courtesy of the Institute for Advanced Study)

A leading advocate of supersymmetry, John Ellis is a theorist at CERN who has focused, among other areas, on the physics that can be done with new particle colliders. (Photograph courtesy of CERN)

Joseph Lykken is a theoretical physicist at Fermilab whose research interests include "supersymmetry, the Higgs boson, and beyond." (Photograph courtesy of Fermilab)

Martinus Veltman (depicted here) shared the 1999 Nobel Prize in Physics with his former student Gerard 't Hooft "for elucidating the quantum structure of electroweak interactions." (Photograph courtesy of CERN)

"We don't know what to expect," says Harvard theorist Matthew Schwartz regarding what the next-generation accelerator might turn up. "But we'll never know if we don't build the machine. And once we do know, we won't have to look again." [57]

Martinus Veltman—co-winner of 1999 Nobel Physics Prize for the contributions that he and Gerard 't Hooft made to the Standard Model—doesn't see an alternative to such a course of action because he believes that investigations at the subatomic level should never end, and most likely *will* never end. "Exploration is part of human activity," Veltman maintains. "Suppose that we decided not to build any more particle colliders, and 50 years from now we're sitting there in a room staring at each other. Don't you think that at a certain point someone will say: shouldn't we at least have a look?" [58]

This kind of approach has always paid off before, says Arkani-Hamed. "In the past century, we've learned something profound about the way the world works in every factor of ten we've taken to shorter distances. There's absolutely no reason to expect that that's going to stop now." To him, the future is clear: We need to build the next big accelerator, colliding protons at energies up to ten times higher than can be done by the LHC, and probing shorter distances than we've ever been able to probe before. [59]

Seeing no new particles after all that effort would come as a shock, as it would suggest there is no simple and "natural" explanation for the fact that the Higgs mass appears to be so artificially low. "The notion of naturalness has been a guideline in physics over the past couple of centuries," claims Nathan Seiberg of the Institute of Advanced Studies. "If naturalness fails, we might have to accept the idea that the parameters are what they are," offering us no obvious rationale as to how or why they have the values they do. Although he hopes that naturalness prevails, seeing it fail would be a momentous occasion—one that would inevitably send theorists back to their drawing boards. [60]

A null finding like this would also be a blow to the reductionist approach, the idea that if there's something we don't understand at a certain length scale, we might make sense of it when we look at smaller distance scale. Reductionism and naturalism have steered physics towards success for hundreds of years, and a major paradigm shift would be called for if they were shown to be wanting. [61]

Right now, we can only guess as to which path nature has chosen. But we have identified the strategy most likely to give us an answer, and it happens to be fairly obvious: It involves building a big machine, perhaps on the order of 100 TeV, which is, of course, bigger than anything we've assembled before. That will take some time—at least a couple of decades of research and development, planning, and eventually construction. There is little advantage in waiting and much to be gained in starting that process soon.

Chapter 4

China on Center Stage

THE IDEA OF EXPLORING NEW PHYSICS WITH THE "GREAT COLLIDER" is, at this moment, just that—an idea whose implementation, even under the most hopeful circumstances, is well over a decade off. Although China has lagged behind the Western World in almost every area of physics—in some cases decades behind—progress is being made to close the gap. Of course, that won't happen overnight. In the meantime, Chinese researchers, based at home and abroad, are working hard to uncover new principles—and new particles as well—that come into play beyond the Standard Model. And they are pursuing this goal in a variety of ways from a number of far-flung locales.

Take the South Pole, for instance, where Chao-Lin Kuo is helping to spearhead the search for gravitational waves from the Big Bang. Kuo, a Stanford physicist who's originally from Taiwan, is a principal investigator in the Background Imaging of Cosmic Extragalactic Polarization 3 experiment based at the Amundsen-Scott South Pole Station in Antarctica. Kuo headed the group at Stanford that played a key role in developing the BICEP3 telescope and detector, and he also designed the detector for BICEP3's predecessor, BICEP2. In the spring of 2014, Kuo and the other lead investigators of BICEP2 thought they had identified signs of gravitational waves produced within a fraction of a second of our universe's birth. The signal they detected took the form of a specific pattern, which had been predicted more than 15 years earlier, in the vestigial light from the Big Bang known as the cosmic microwave background (CMB).

If true, this would have been the first direct evidence supporting the idea of cosmic inflation—a brief and explosive expansionary epoch that is believed to have provided the driving force, or kick, behind the Big Bang. However, data subsequently released by the Planck space telescope—and jointly analyzed by Planck and BICEP2 scientists—showed that the effect seen by the South Pole telescope was largely the result of dust in our galaxy rather than being of primordial origin.

The new telescope, BICEP3, is making observations in different wavelengths that are less susceptible to the influence of dust, thus having a better chance of witnessing phenomena that arose during the universe's birth throes—and at energy scales a trillion times higher than manmade accelerators like the LHC can reach. The recently acquired potential to detect gravitational waves, which owes in large part to the technology developed by Kuo and his BICEP collaborators, opens up a new realm to explore and, along with it, says Kuo, "prospects for new discoveries. It gives researchers hope that new physics"—representative of conditions that prevailed at the dawn of time—"might finally be coming within our reach." [1]

At the same time, some members of the BICEP team are looking for a northern hemisphere site from which to make complementary CMB measurements, and they think they may have found it in the Ali region of western Tibet. The person leading this effort is Meng Su, an astrophysicist based at MIT and IHEP, as well as a BICEP collaborator, who is originally from Taiyuan, China. With observations taken from Antarctica, from Chile (at the Atacama Cosmology Telescope), and from Tibet, says Su, "you can achieve full coverage of the CMB sky from ground telescopes. If you detect a signal from Antarctica that appears to come from the primordial universe and detect a similar signal while looking in the opposite direction from Tibet, using different instruments, you can be much more confident about making a claim regarding the discovery of inflation and gravitational waves. This could provide an important cross-check to what would constitute an extremely momentous finding." [2]

The Ali region is an appealing locale for a CMB observatory because of its elevation—5100 meters and higher in this region, which leaves less atmosphere to look through—and its exceedingly dry conditions, owing to the fact the Himalaya Mountains trap water vapor and keep it from reaching the Tibetan plateau. These factors should combine to make for what astronomers call "good seeing." Several small optical telescopes, owned and operated by the National Astronomical Observatory in China (NAOC), already exist at the site, which means that the necessary infrastructure—in terms of roads, electricity, and support buildings—

are in place. Without having to engage in extensive site development, an experiment could be carried out at relatively low cost, on the order of $10 to 20 million.

IHEP is supporting the project and may provide some seed money, but the Institute, most likely, will not be able to subsidize the entire experiment by itself. Before embarking on a fundraising campaign, Su plans to travel to Ali before the end of 2015, accompanied by several CMB authorities—including John Kovac of Harvard and Clem Pryke of the University of Minnesota, principal investigators of the BICEP experiment, and John Carlstrom of the University of Chicago. The goal of this outing is simple: to find out how good the Tibetan site really is for CMB astronomy. And if it's as good as good as these experts think it may be, Su will kick into high gear, doing everything in his power—and taking advantage of every connection he has—to enable this experiment, of potentially cosmic significance, to take place. [3]

While Tibet is the world's highest and largest plateau, sometimes referred to as "the roof of the world," there is also interesting science to be done at much lower elevations. The Jinping Deep Underground Laboratory, buried beneath 2,400 meters of rock in a remote section of China's Sichuan Province, has been dubbed the deepest laboratory in the world" by the journal *Nature*. Jinping is home to the Particle and Astrophysical Xenon (PandaX) experiment, which is designed to look for dark matter from the inner recesses of the earth. [4] The laboratory's great depth—and the fact that it's encased in a deposit of solid marble— offers better shielding from background noise than similar dark matter searches operating in Italy's Gran Sasso Tunnel (the Xenon100 experiment) and in a former South Dakota gold mine (the Large Underground Xenon, or LUX, experiment). The PandaX collaboration was initiated in 2010, before the laboratory was opened, and has since grown to include some 40 scientists, mostly from China, with a few participants from the United States. "We started off basically from nothing," says Xiangdong Ji, a physicist at Shanghai Jiao Tong University who serves as the group's spokesperson. "We didn't have a group, we didn't have equipment, we didn't have anything." But as *Science* magazine reported in 2014, "the PandaX team is quickly catching up to the rest of the world." [5]

Like other searches of its kind, PandaX aims to detect and eventually characterize the dark matter that is believed to comprise more than 84 percent of the universe's matter. "Although ... indirect astrophysical observations convince us that dark matter exists, dark matter has not yet been directly observed," the collaborators wrote in a paper that described the detector's performance during the commissioning phase. "The Standard Model of particle physics, which has been very successful in explaining the properties of ordinary matter, can neither explain dark matter's existence nor its properties." [6]

PandaX, Xenon100, and LUX all seek to find a new class of particles called WIMPs, short for weakly interacting massive particles, which is the form that many physicists believe dark matter assumes. A common premise underlying such efforts is that our galaxy could be awash in particles that have eluded discovery so far simply because they almost never interact with anything else. One way to snare such a particle might be to set a "trap" for it by placing a reservoir of an especially dense material deep underground—where it will be sheltered from background radiation and other potentially confusing signals—in the hopes that dark matter will, upon occasion, collide with a target atom. The favorite target at the moment consists of liquid xenon—cryogenically cooled to temperatures of about -100C—housed deep underground[7]. The xenon is kept cold so that the atoms inside the detector are barely moving, making it much easier to notice if and when an atom gets a sudden jolt from a collision with an invisible dark matter particle. On the rare instance of a WIMP ramming into a xenon nucleus, a flash of light would be emitted as the nucleus is sent hurtling through the dense medium, some three times thicker than water. Electrons from the stricken atom would also be liberated during the crash, giving rise to secondary bursts of light, which could also be spotted by sensors that surround the detector.

One advantage of PandaX, according to Ji, is that it can be scaled up rapidly. In fact, he and his collaborators are soon planning to implement a 20-fold expansion of the xenon detector from its initial size of 120 kilograms[8] up to 2,400 kilograms for the second round of experiments, which would significantly boost the instrument's chances of observing WIMP collisions.[9] If funding for this upgrade comes through, Ji will be well on his way toward realizing one of the original goals he and his colleagues have set for this venture. "We want to demonstrate that world-class research in dark matter is possible in China," he says.[10] An even greater outcome, of course, would be to catch dark matter in the act of flying through its xenon-filled trap—a discovery, the PandaX collaborators say, that would simultaneously have "a profound impact on cosmology, astronomy, and particle physics."[11]

Meanwhile, a parallel experiment is underway in space, circling the Earth every 90 minutes at an altitude of about 350 kilometers. The Alpha Magnetic Spectrometer (AMS) is being carried on the International Space Station to sift through cosmic rays—high-energy particles, spewed out of exploding stars and other sources—in the hopes of learning about dark matter and perhaps revealing other exotic physics as well. This huge collaborative effort, involving some 600 researchers from 56 institutions and 16 countries, is headed by Samuel Ting, an MIT physicist and Nobel laureate. Ting grew up in mainland China and Taiwan where—because of an ongoing war between China and Japan in the late-1930s and 1940s and because of his family's frequent moves—he had almost no formal schooling as a youth, instead learning on his own about great scientists like Isaac Newton, Michael Faraday, and James Clerk Maxwell.[12]

Ting started pursuing the AMS in 1994, one year after the Superconducting Super Collider (SSC) was cancelled. If we could not build a collider powerful enough to produce dark matter particles, he surmised, perhaps we could go up to *space,* where these particles supposedly abound, and study them there, away from the confounding influences of Earth. The AMS is similar, in principle, to the detectors typically used in particle accelerators. It has a magnet that bends particles that fly into it—coming from cosmic rays, in this case, rather than from artificially induced collisions. The degree of bending depends on the particle's mass, and the direction that a particle bends (as discussed in earlier chapters) depends on its charge. Tracking devices provide clues about a particle's velocity, mass, and charge; and by pooling that information, an identification can hopefully be made. The AMS is also similar, in some respects, to experiments like PandaX, but its measurements are done in outer space to shield the experiment from terrestrial interference. From its lofty perch, 400 or so kilometers above the Earth, the AMS gets to examine pristine cosmic rays that are not distorted by the atmosphere, in some cases having traveled through space, unperturbed, for millions of years.

The 7.5-ton magnetic spectrometer reportedly cost somewhere between $1.5 and $2 billion to build, from funds that Ting raised largely on his own. The weighty device was carried to the space station in May 2011 by the Space Shuttle Endeavor on its last voyage ever. The launch occurred when Ting was 75 years old—17 years after he had started working on the project—but this liftoff would never have happened were it not for his iron will and forceful diplomacy. Ting is, by most accounts, a driven personality who cannot easily be deterred from his goals, once telling the U.S. Department of Energy (after one of his proposals had been turned down): "I reject your rejection." [13]

A space shuttle flight dedicated to the AMS in 2005 was cancelled after the Columbia Space Shuttle blew up in 2003, killing all seven crew members. It took tireless lobbying on Ting's part, and a vote by both houses of the U.S. Congress, to authorize an additional space shuttle flight that would put the AMS in orbit before the shuttle program was terminated for good. "Without [Ting's] absolute unwillingness to give up, we would not have gotten it," says former U.S. Senator Kay Bailey Hutchison, who helped secure funding for the project. [14]

Shortly before the instrument's scheduled launch in 2010, Ting decided to remove the powerful superconducting magnet at the center of the device and replace it with a permanent magnet that would be somewhat weaker yet would enable the experiment to keep running many years longer. That last-minute switch caused an additional delay, resulting in another missed flight, but the AMS finally lifted off from the Kennedy Space Center on May 16, 2011. It was mounted to the International Space Station a couple of days later and started collecting data shortly thereafter.

By 2014, the instrument had gathered information on 54 billion cosmic ray events, and 41 billion particles were picked out for more detailed scrutiny. [15] The AMS analyzes incoming cosmic rays with unprecedented accuracy, acquiring data at energies never before studied—cataloguing particles up to 500 GeV so far, [16] with plans to extend that analysis in coming years up to 1,000 GeV (or, equivalently, 1 TeV). [17]

Researchers are particularly interested in charting the relative numbers of electrons and positrons collected at different energy levels. The rationale for this strategy stems from the fact that many theoretical models (including most of those involving supersymmetry) hold that a dark matter particle is its own anti-particle. Therefore, when two dark matter particles, or WIMPs, collide they will annihilate or disappear, producing in their stead other particles such as photons, electrons and positrons, or protons and antiprotons. The particles produced from WIMP annihilation would look like normal cosmic rays, except that they would appear at higher energies and have different relative abundances.

Seeing a rise in the ratio of positrons to electrons at a certain energy, and a sudden drop-off above that energy level, could be a "smoking gun signature" of dark matter collisions. "The way they [positrons] drop off tells you everything," Ting notes. [18] The AMS team saw evidence of a decline in the number of positrons above 275 GeV, but they need more detailed measurements to determine the rate of decrease. [19] That could provide clues as to whether the positrons in question were produced by dark matter or more likely came from known astrophysical sources, such as emissions from rotating neutron stars called pulsars. It's a possible indication of dark matter, says Ting, "but by no means is it proof." [20]

An April 2015 announcement by the AMS team discussed measurements of the antiproton to proton ratio observed in cosmic rays across a wide energy band. These results, like those found for the positron-electron ratio, also appeared to be at odds with what would be expected from ordinary cosmic rays, but there was insufficient data to definitively attribute the findings to dark matter. [21] A clearer picture should come with the accumulation of more data, and if all goes well the experiment could keep running through 2020 and maybe until 2028. [22] No one knows what will turn up in the end because we are opening a door into a new domain that hasn't been carefully explored, says Ting, a meticulous investigator who closely oversees every aspect of his experiments. "My most important responsibility is to make sure the instrument is correct. We don't know what we'll see, but you have to be sure that what you see is correct." [23]

While conclusive evidence from the AMS has yet to come in, important results have already been obtained at a separate experiment taking place near Daya Bay in southeastern China, some 55 kilometers northeast of Hong Kong. Neutrinos, the focus of this experiment, come in three varieties or "flavors"—electron, muon, and tau—which are associated through various decay processes with the three other (charged) leptons in the Standard Model, the electron, muon, and tau. An international group of researchers at Daya Bay, led by the Chinese and U.S. contingents, are exploring a number of mysteries about neutrinos, including their masses (which have yet to be accurately pinned down), plus their uncanny ability to transform from one type into another.

Neutrinos—phantom creatures produced in fantastic numbers during the Big Bang—are, after photons, the second most abundant particles in the universe, outnumbering all other forms of matter. [24] They travel through space at close to the speed of light, passing through anything and everything in their path, while rarely skipping a beat. "About 100 billion neutrinos from the sun pass through your thumbnail every second, but you do not feel them because they interact so rarely and so weakly with matter," said John Bahcall, the late astrophysicist who was formerly based at Princeton's Institute for Advanced Study. If all of these solar neutrinos were to continue on their downward trajectory and pass through the Earth, only about one of the 100 billion would interact in any way with the material out of which the Earth is made. [25]

The original version of the Standard Model held that neutrinos were massless, but evidence turned up since the late 1990s has shown that these particles are endowed with a modest amount of mass, perhaps on the order of one millionth that of an electron—findings that have posed a challenge for theorists who are looking for the best way of modifying, or "extending," the model. "Whatever more satisfying theory turns out to be the next step beyond the Standard Model," Steven Weinberg said, "this theory is likely to entail the existence of small neutrino masses." [26]

Although it is not known whether neutrinos come by their mass through the Higgs mechanism or by some other means, the existence of neutrino mass is closely linked to the particle's ability to regularly change its form. This phenomenon was documented in 2001 as a solution to the "solar neutrino problem"—the puzzling fact that only one third as many electron neutrinos that were predicted to reach the Earth from the sun were actually seen. The discovery was made at the Solar Neutrino Observatory in Sudbury of Ontario, Canada. "The solution of the mystery of the missing solar neutrons is that neutrinos are not, in fact, missing," Bahcall explained. "The previously uncounted neutrinos are changed from electron neutrinos into muon and tau neutrinos that are more difficult to detect." [27]

Neutrino oscillation is a quantum mechanical phenomenon—a consequence of the premise that a neutrino starts out with a specific flavor but is not a discrete, immutable object. Instead it contains a combination of different mass states, and over time it will inevitably switch from one variety of neutrino to another and back again. As an analogy, Ming-Chung Chu—a physicist at the Chinese University of Hong Kong and a Daya Bay collaborator—suggests a slowly rotating coin.

"Sometimes you see only heads, sometimes only tails, and sometimes a little bit of both," Chu says. "That is like neutrino oscillation, although an even better comparison would be with a three-sided coin," assuming that such an object could be found." [28]

Harvard's Matt Strassler suggests another illustration: A road heading in a northeasterly direction, he says, "is both partly a north road and partly an east road. You cannot say it is one or the other. And a northward-tending road is a mixture of a northeast road and a northwest road. And so it is with neutrinos." Neutrinos of each of the three flavors contain "a precise but different mixture" of the three possible mass states. Rather than being a static thing, this mixture changes dynamically, giving rise to periodic shifts in identity. [29]

The length of time between oscillations, or the period, depends on the mass difference between flavors. As the mass difference between flavors gets smaller, the period gets longer. If there were no mass difference at all, the period would become infinitely long and oscillations, consequently, would not occur at all. The fact that oscillations have been seen to occur among all three flavors shows that electron, muon, and tau neutrinos must have different masses. If that is the case, no more than one of these particles, at the very most, can have zero mass. There-fore, at least two and perhaps all three flavors of neutrinos are massive—a finding that is at odds with the Standard Model as originally conceived.

Because neutrinos are so light, physicists have not yet come up with a way to directly measure their mass. But we can still learn about their mass through neu-trino oscillation measurements. Daya Bay investigators are particularly interested in measuring the probability of a neutrino of one flavor transforming into an-other—a parameter known as the *mixing angle*. Two mixing angles had already been established, quantifying the rate at which two sorts of transitions occurred. But a third mixing angle—called "theta one-three," which concerns an electron neutrino morphing into either a muon or tau neutrino—had not been measured definitively before, nor had it been shown to be nonzero. "A vital open question was whether the three types of neutrinos genuinely mix three ways or whether what we were seeing was just two separate pair-wise mixings," notes Jon Butter-worth, who heads the physics department at University College London. [30] And that is the very question that the team at Daya Bay took on.

The site was chosen because of its proximity to a complex of six nuclear reac-tors—from the Daya Bay and Ling Ao power stations—that churn out prodigious quantities of antineutrinos at the rate of more than a billion trillion a second. [31] Antineutrinos, which have the same mass, the same (neutral) charge, and the same spin as neutrinos, can be used in experiments to address the same ques-tions about oscillations and mixing angles that physicists have posed about neu-trinos themselves. (There is, however, a chance that neutrinos and antineutrinos

The Alpha Magnetic Spectrometer-02, a project conceived of and initiated by Samuel Ting, is shown attached to the International Space Station. (Photograph courtesy of NASA)

A view of the underground antineutrino detectors at the Daya Bay Reactor Neutrino Experiment (Photograph courtesy of IHEP)

IHEP director Yifang Wang at the site of the Daya Bay experiment, where he serves as co-spokesperson (Photograph courtesy of IHEP)

oscillate somewhat differently, and that is a possibility that physicists are eager to investigate.)

To prepare the neutrino—or, rather, antineutrino—laboratory, more than three kilometers of tunnels and three large cavernous halls, needed to house eight neutrino detectors in all, were blasted out of a rocky hillside. The excavation was completed in late 2010, and the detectors were installed in the following year beneath an overburden of several hundred meters of granite, constituting an effective barrier against cosmic rays. Each antineutrino detector is a cylinder, weighing 110 tons, which has a tank in its center filled with 20 tons of liquid "scintillator" containing gadolinium, a heavy metal. When an antineutrino interacts with a proton in this liquid, a positron and neutron are produced, and the energy from this reaction creates a burst of light. Photomultiplier tubes, which line the detector's outer rim, record the light flashes that accompany an antineutrino sighting. The detectors, in turn, are submerged in pools of pure water to provide further protection from cosmic rays as well as shielding from the radioactive decay occurring in the surrounding rock.

Two of the halls, which can accommodate two detectors each, are located about 250 meters from the nuclear plants, and another hall, hosting four additional detectors, is located about two kilometers from the reactor cores—the distance at which the probability of an electron antineutrino "disappearing" reaches a theoretical maximum. The strategy adopted by the researchers is fairly straightforward: Their detectors are only sensitive to electron antineutrinos and cannot see muon or tau antineutrinos. Therefore, they counted the number of electron antineutrinos recorded at the first set of detectors, close to the reactor source, and the number recorded at the more remote detectors. They then calculated the number that would have been recorded at the "far site" had there been no antineutrino oscillations. With these numbers in hand, the Daya Bay physicists determined the number of electron antineutrinos that converted to another form that is invisible to their detectors. From that, they were able to compute the third, and possibly final, mixing angle with far greater precision than had ever been achieved before.

"We were ten times better than previous experiments in reducing systematic error," says IHEP's Yifang Wang, who co-heads the project with Kam-Biu Luk, a physicist based at the University of California, Berkeley and the Lawrence Berkeley National Laboratory.[32] The Luk- and Wang-led collaboration reported its findings on March 8, 2012, with the first statistically significant determination of theta one-three—a measurement that relied on six of the eight detectors eventually installed at the underground facility. The finding was nearly as solid as the granite layer in which the experiment was embedded. It had an accuracy, or confidence level, of 5.2 sigma, meaning that the odds of the result being a fluke were quite slim—one in several million. Corroboration came one month later when the Reactor Experiment for Neutrino Oscillation (RENO) in South Korea announced a similar result.

The mixing angle (theta one-three) identified by the Daya Bay experiment was not only nonzero—showing that three-way mixing among neutrinos actually occurred—it also had a bigger value than expected, which was good news for physicists who hope that neutrinos might shed some light on one of the greatest mysteries of all: why our universe, which presumably started out with equal amounts of matter and antimatter, now appears to be almost exclusively matter.

The explanation may be due, at least in part, to subtle differences in behavior between matter and antimatter. Such differences, should they arise, would constitute a violation of charge parity (CP) symmetry—an assumed symmetry between matter and antimatter, which holds that matter and antimatter particles are interchangeable and can be substituted for each other in various processes without affecting physical outcomes.

The first hard evidence of a CP violation came in 1964 from experiments at Brookhaven National Laboratory carried out by James Cronin and Val Fitch. Cronin and Fitch looked at kaons, also called K-mesons, which are short-lived pairings of quarks and antiquarks. The neutral variety of kaons the duo focused on have a strange property: they are constantly shifting between their matter and antimatter forms. Cronin and Fitch observed, however, that these transformations do not occur with equal probability in both directions—a finding that earned them the 1980 Nobel Prize. The kind of CP violation that they witnessed is called "indirect" because it related to the way the two kaon states mixed with each other or fluctuated back and forth—processes that are governed by the weak force.

A direct CP violation would relate to the actual decay of particles such as kaons (which are also weak force processes), rather than the state that the decaying particles are in. This is a much more feeble effect and therefore much harder to detect. It was finally seen in 1999 by the KTeV experiment at Fermilab's Tevatron collider and the NA48 experiment at CERN's Super Proton Synchrotron, both of which scrutinized the decay of neutral kaons. When these particles decay, they leave behind a charged pion, a neutrino, and either an electron or its antiparticle, a positron. If CP symmetry were preserved, electrons and positrons would be produced in equal numbers. But the researchers saw something different: CP symmetry was violated, there being slightly more electrons turned out than positrons.

If a similar process occurred in the early universe, the electrons and positrons would annihilate each other, leaving some electrons still in circulation. The magnitude of the surplus of electrons over positrons, however, would be far too small to explain the imbalance that exists between matter and antimatter in the universe at large. That's why scientists are interested in seeing other examples of CP violations, involving other particles, including those that relate to ubiquitous neutrinos, where the effects may well be larger.

The kind of deviations from CP symmetry that neutrino physicists are looking for would involve slight differences in the behavior of neutrinos and antineutrinos. The Daya Bay result was encouraging, explains Paul Langacker of the Institute of Advanced Studies in Princeton, because "all these CP-violating effects vanish if theta one-three is zero." [33] And, fortunately, the mixing angle had a large enough value to make researchers think that CP violations in the neutrino sector, if they exist, may be observable.

Follow-up experiments are already planned to see whether neutrinos oscillate differently than antineutrinos and if there's a big enough difference to account for the disparity in prevalence between matter and antimatter. Now that we know that theta one-three is nonzero, says Luk, "we can go forward and hunt for CP violations." [34]

Work at Daya Bay continues with all eight detectors in full operation, which means that scientists can get even more accurate measurements of the theta one-three mixing angle. In August 2013, the team announced a new finding—a precise measurement of one of the "mass splits," or mass differences, between two types of neutrinos. "In principle," says Luk, "if we can measure all of the differences between the masses [of the three neutrino flavors] to very high precision, we should be able to deduce their order. This is a step toward solving the problem of the mass hierarchy," which involves figuring out the relative masses of the three neutrino types. [35]

As a result of Daya Bay, says Ming-Chung Chu, summing things up, "we know all three neutrino mixing angles, and we also know that all three neutrinos have different masses. We further know that two of those masses are pretty close to each other." [36] The next step is to work out the mass hierarchy, or ordering of neutrino masses, without working out the absolute value of those masses, which, unfortunately, is not yet within reach. And that's the goal of the Jiangmen Underground Neutrino Observatory (JUNO) experiment that Yifang Wang is now heading up.

In January 2015, construction of the JUNO detector was initiated near the city of Jiangmen in China's Guangdong Province, about 100 kilometers west of Hong Kong. Almost everything about this venture will be bigger than Daya Bay. The success of the latter effort, says Wang, has attracted more foreign partners for JUNO, with more than 300 collaborating scientists already on board. [37] JUNO's detector, consisting of 20,000 tons of liquid scintillator—more than one hundred times bigger than all of Daya Bay's detectors put together—is expected to be the largest and highest precision of its kind in the world. The experimental apparatus, encased within a 38-meter-diameter sphere, will be buried inside a hill beneath almost 730 meters of granite rock, placed midway between two nuclear power complexes that are roughly 50 kilometers away in either direction.

Data taking should begin by 2020 and, within six years, Wang believes they should have a determination of the mass hierarchy of neutrino types—to an accuracy of at least 4 sigma (which would still fall short of the discovery threshold). [38] While seven other experiments—in Asia, Europe, and the United States—are going after the mass hierarchy measurement, Wang is confident that JUNO holds the edge. "With leading liquid scintillator detector technologies, rich experience in reactor neutrino experiments, and innovative experimental design," he says, "the JUNO experiment could be the first to determine the neutrino mass hierarchy." [39] Such an accomplishment would mark a new beginning, rather than an endpoint, he adds. "Once we figure out the mass hierarchy, we might be able to figure out what modification is needed so that the Standard Model can accommodate neutrino masses." [40]

JUNO also has a chance of contributing in a different—yet related—area. By making more precise measurements of the neutrino mixing angles and mass differences, the experiment could turn up the possibility of a fourth flavor of neutrinos—a finding that would require further modifications of the Standard Model and possibly overturn that model altogether. However things turn out, Wang is confident that the endeavor will be a boon for Chinese science in general. "JUNO will help us build a leading research team, and make China one of the leaders in the field of particle physics," [41] he says.

Even while Wang looks ahead to JUNO, which will attempt to go past Daya Bay on multiple fronts, he is mindful of—and grateful for—the former experiment's success. Up until now, he says, the Daya Bay experiment has been one of the largest international collaborations in basic science in China and the biggest U.S.-China collaboration—an effort that has involved six countries (China, the Czech Republic, Hong Kong, Russia, Taiwan, and the United States), 38 universities, and more than 250 scientists. [42]

Fortunately, this cooperative effort has paid off. "This [mixing angle measurement] is arguably the most important physics result ever to come out of China," says Robert McKeown, a Daya Bay collaborator and deputy director of the Thomas Jefferson National Accelerator Facility in Virginia. [43] Hervé de Kerret, a physicist who directs the competing Double Chooz neutrino experiment in France, agrees, calling the Daya Bay finding "a perfect confirmation and a beautiful result." [44]

The determination of the mixing angle, as presented in a paper submitted to *Physical Review Letters* in March 2012, was named one of the top ten breakthroughs of the year by *Science* magazine. "Chinese particle physics has arrived, it seems," *Science* reported. [45] The result, the magazine added, put the nation's particle physics program "on the map." [46] While it was an admittedly important milestone, marking China's entry into the world of cutting-edge, basic research, Wang stresses that the goals of the Chinese physics program are far more ambitious, extending well beyond a single result. [47]

In 2014, he and Luk were named co-winners of the 2014 W.K.H. Panofsky Prize in Experimental Particle Physics, named after Wolfgang Panofsky, the former director of the Stanford Linear Accelerator Center (SLAC). The American Physical Society awarded the 2014 prize to Luk and Wang "for their leadership of the Daya Bay experiment, which produced the first definitive measurement of the theta one-three angle of ... neutrino mixing." [48]

Daya Bay, says Chu, "is a major international collaboration in China, an important precedent and a necessary first step before considering a project on the scale of the Great Collider, which would necessarily involve thousands rather than hundreds of people." [49]

Before getting to that project, which is, of course, the centerpiece of this book, we'll first say something about how experimental particle physics took hold in China, where the field stands today, and the extent to which the groundwork has been laid for what many hope to be a giant leap into the future. This story, as with the narrative concerning Daya Bay, involves critical collaborations between China, the United States, and other nations. Influential roles were played in particular by the Chinese-born Nobel laureates in physics, Tsung-Dao (T.D.) Lee, Chen Ning Yang, and Samuel Ting, as well as by the aforementioned Wolfgang Panofsky.

Although both Lee and Yang grew up in China, they completed their doctoral studies in the United States, where they also pursued their work on the violation of parity conservation that won them the 1957 Nobel Prize. Samuel Ting was awarded his Nobel Prize in 1976 for his co-discovery of the J/Psi particle based on research that, again, was conducted in the United States (see Chapter 1). More Nobel Prizes could start coming out of China itself with the establishment of a "tradition of genuine science culture," says Gerard 't Hooft, himself a Nobel laureate, but that takes time. [50]

Given that this sort of homegrown scientific culture had not been established earlier in the 20th century, and is still not there yet, other Chinese physicists— who arrived on the scene before Lee, Yang, and Ting—similarly made some of their most noteworthy contributions to the field while they were off in foreign lands. The principle reason for these excursions was that Chinese universities and laboratories could not match the leading facilities in Europe and the United States, having neither the technology, technical expertise, nor the academic infrastructure necessary to support world-class research.

Wu Youxun, for example, verified the Compton effect—the transfer of energy from a photon to an electron during a collision between these particles—as a graduate student at the University of Chicago in the 1920s. Wu's experiment helped his advisor, Arthur Compton—who first observed the effect—win the Nobel Prize in 1927. In a paper published in 1930, the Chinese physicist Zhao Zhongyao described experiments he'd completed in the previous year as a graduate student at the California Institute of Technology. In this work, Zhao observed a phenomenon later confirmed to be electron-positron annihilation. His classmate at Caltech, Carl Anderson, acknowledged that Zhao's research had inspired his 1932 discovery of the first antimatter particle, the positron—a finding that earned Anderson the Nobel Prize in 1936.[51] In 1959, the Chinese physicist Wang Ganchang was working at the Joint Institute of Nuclear Research in Dubna, USSR, when he and his colleagues discovered the anti-sigma minus hyperon, the first electrically charged antihyperon whose existence had been anticipated by theorists. Wang's finding affirmed the notion that a hyperon, which is a baryon with one or more strange quarks, had a corresponding antiparticle. In fact, Paul Dirac's theory, first presented in 1928, predicted the existence of the positron, while suggesting that any subatomic particle of spin-1/2 should have an associated antiparticle. Given that the sigma minus hyperon—the matter counterpart of the anti-sigma minus hyperon—is a spin-1/2 particle, Dirac's broader prediction was also upheld in this case.

Wang was a cofounder of Dubna's Joint Institute, where a number of Chinese physicists came to gain experience with accelerator design and experimentation. But he longed for the day when he and his compatriots would be able to carry out research of this sort in China without having to travel abroad—an idea that finally started gaining some forward momentum in the early 1970s.

On March 15, 1971, the U.S. State Department eased restrictions on Americans with passports who wished to go to China. Chen Ning Yang was one of the first Chinese-American scientists to make that trip once the travel ban was lifted. During his visit, Yang met with China's longstanding premier, Zhou Enlai. Yang told Zhou that scientific training in the country was particularly lacking in both theory and basic research, and he continued to push these themes during a subsequent visit in 1972. Zhou passed on these suggestions in a conversation with Chairman Mao Zedong, while urging the research institutes at the Chinese Academy of Sciences (CAS) to focus more on basic science.

Mao was receptive to these ideas, given his personal interest in particle physics. In a discussion with philosophers on the implications of new theories about elementary particles, he asserted in 1964 that "matter is infinitely divisible,"[52] making the separate claim that the electron, like the atom, would ultimately be split.[53] The converse of that statement was hard for him to fathom. For if matter could not be divided indefinitely, Mao added, and "if there is an end, there is no science."[54]

Meeting with Zhou in October 1972, Tsung-Dao Lee called for the formation of a high-energy physics program in China. Zhou should invite foreign scientists to visit the country, Lee advised, and he should also send Chinese students and scientists to train at research centers in Western countries—at places like CERN and the U.S. national laboratories—where he was certain they would be welcome. [55] Zhou agreed, stating in a letter to the physicists Zhang Wenyu and Zhu Guangya, "This issue should not be delayed any further. The study of high-energy physics and the research and development of high-energy accelerators should be one of the main projects of CAS." [56] Zhou also called on Zhou Peiyuan—the vice president of Beijing University and the Chinese Academy of Sciences who had previously studied general relativity under Albert Einstein—to encourage more basic physics research at his university and throughout the country at large.

The visits by Yang and Lee, who became national heroes in China upon winning their Nobel Prizes, did much to boost physics programs in their native land and to make the case that top-flight research not only could take place there but should. Yang once said that the most important contribution he'd ever made was "to help the Chinese change their perception that the Chinese were not as talented as others" and to show that they could excel in a field that had been dominated by the Western world for centuries. [57]

Ting was also celebrated among the Chinese people for his Nobel Prize-winning work. In fact, when he accepted his award in Stockholm, Sweden in December 1976, Ting became the first person ever to have addressed the Nobel audience in a Chinese language. In his speech, he stressed a point that he has illustrated, time and again, throughout his illustrious career—namely, that the work of experimentalist physicists is every bit as important as that of theorists, even though the latter are often accorded a higher status in the pecking order. Ting, however, has steadfastly insisted—and demonstrated through his work—that the field needs both experimentalists and theorists to thrive. [58]

With Zhou Enlai's support, the Institute of High Energy Physics (IHEP) was founded in February 1973. Zhang Wenyu—who had taught at Purdue University in the U.S. from 1949 to 1956 before returning to China—was made the founding director, using the position to promote the cause of high-energy physics back home. (Zhang, incidentally, was followed in the director's position by Minghan Ye, Shouxian Fang, and by three former students of Ting's—Zhipeng Zheng, Hesheng Chen, and the current director, Yifang Wang.)

In keeping with Lee's recommendation, Zhou also promoted the idea of an exchange program in high-energy physics. Zhang visited SLAC on his own in 1972, and in the following year he led a delegation of Chinese physicists on a tour of U.S. and European labs. The group spent about two months, from May to July 1973, visiting SLAC, Fermilab, and Brookhaven National Laboratory, stopping off at CERN on the way back. The physicists returned to China enthused about the possibility of building a 50 GeV proton collider, which would have made it among the most powerful accelerators in the world—comparable in energy to that of the SLAC collider itself (although SLAC's machine was accelerating electrons and positrons rather than protons). [59] In March 1975, Zhou endorsed a report of the Chinese Academy of Sciences on the proposed proton collider, and Project 753 was born, named after the year (1975) and the month (3) in which it had received an initial go-ahead.

Lee, however, was not enthusiastic about this proposition, feeling that China's interests would be better served by a lower-energy electron-positron collider that would also be considerably less expensive to build and operate. Panofsky shared Lee's reservations, maintaining that a smaller, electron-positron accelerator on the scale of about 2 GeV would offer China a better entrée into the world of experimental high-energy physics and a better chance of making real contributions to science. Such a machine, Panofsky had argued, "could serve a dual purpose," also providing a source of synchrotron radiation that could be used for materials science research and other laboratory applications—an idea that was ultimately pursued to supplement the fundamental physics goals.

Zhou, however, continued to promote Project 753 until his death in January 1976. Mao Zedong died about five months later, and the project was put on hold until October 1976 when the "Gang of Four"—a faction of Communist Party leaders that were hostile to high-energy physics research and unfavorably disposed towards basic scientific research in general—was finally deposed.

Panofsky had expressed his concerns about the proposed 50 GeV collider with Zhang Wenyu who, as IHEP's head, held a position of influence in the nation's high-energy physics program. Zhang had visited Panofsky in 1973 during his tour of SLAC, and he invited the American to come to China in October 1976. A devastating earthquake had struck the country in July, centered around the city of Tangshan, killing an estimated quarter of a million people. Panofsky offered to postpone his trip in the wake of this widespread disaster, as he recounted in his memoir, "but Zhang firmly maintained that the importance of high-energy physics was such that we should proceed as scheduled." [60]

Although Lee's and Panofsky's views ultimately prevailed, Hua Guofeng, who led China from 1976 to 1978, backed the construction of a 50 GeV proton machine in 1977, on the heels of a vote of confidence expressed at a Chinese Academy of Sciences-sponsored symposium. But an economic downturn soon forced the scrapping of this idea. [61]

As a follow-up to Panofsky's tour of China, the vice premier of the Chinese State Council, Fang Yi, went to the United States to spend time at SLAC, Fermilab, and other U.S. facilities. The purpose of Fang's visit was to chart out the best path for China to take in starting a high-energy physics program of its own. Fermilab director Robert Wilson supported the recommendation of Lee and Panofsky that a more modest electron-positron facility would be the way to go.

Kam-Biu Luk of the University of California, Berkeley, and the Lawrence Berkeley National Laboratory, as seen standing in front of the Ling Ao Nuclear Power Plant—one of the three nuclear plants used in the Daya Bay neutrino experiment. Luk serves as co-spokesperson for this experiment. (Photograph courtesy of Kam-Biu Luk)

The Institute of High Energy Physics, based on the campus of the Chinese Academy of Sciences in Beijing (Photograph courtesy of IHEP)

A portrait of Zhang Wenyu, the first director of the Institute of High Energy Physics in Beijing (Photograph courtesy of IHEP)

A portrait of the physicist Wolfgang Panofsky, a former director of the Stanford Linear Accelerator Center or SLAC (Photograph courtesy of the Stanford Historical Photograph Collection [SC1071], Department of Special Collections and University Archives, Stanford University, California)

In January 1979, shortly after his return to China, Fang and U.S. Energy Secretary James Schlesinger signed a formal agreement, calling for cooperation between the two nations in the field of high-energy physics. The physics accord was part of a broader Agreement on Cooperation in Science and Technology, which was signed by China's supreme leader Deng Xiaoping and U.S. president Jimmy Carter. Consultations between the two countries ensued, as reported by Panofsky, who attended the signing ceremony. In light of these discussions, the Chinese government agreed to sponsor the construction of the Beijing Electron Positron Collider (BEPC)—an accelerator with an energy of up to about 2.2 GeV per beam, which is what Panofsky and Lee, among others, had been calling for all along. [62]

After reviewing the plan as formulated by the Chinese Academy of Sciences, Deng Xiaoping said: "I am in favor. Do not hesitate." [63] And in keeping with that resolve, things started moving quickly.

Lee recommended that Panofsky, with his wealth of experience in accelerator physics, be named the project's scientific advisor. Panofsky agreed but refused to take any financial compensation from China in order to maintain his independence. In 1982, a group of about 30 Chinese physicists and engineers came to SLAC, where Panofsky was still director, to draw up a preliminary design of the BEPC—a task that was completed before the end of the summer. "It was quite a scene," Panofsky described. "We had 30 Chinese engineers in Mao suits running in and out of our lab." It took a remarkably short time, just four years, to design and mostly build the collider—an effort that involved an extensive collaboration between China and the U.S., particularly SLAC. "[The project] was finished on time and on budget," remarked Hesheng Chen, who became IHEP's director in the late 1990s. [64]

A groundbreaking ceremony, marking the start of construction, was held on October 7, 1984 at a site on the western edge of Beijing. Deng, who attended the proceedings, expressed utter confidence in the undertaking, stating, "I am sure this matter cannot be wrong."

At the time, many people from the U.S. and Europe visited, including the physicist John Adams, a former director of CERN. In a meeting with Deng, Adams expressed some doubts about the undertaking, asking him why China needed this high-energy physics project, especially given that the country's economic base was still rather weak. "We need this for the future," Deng replied. [65]

"This was the biggest scientific project in China," recalled Chuang Zhang, a former deputy director of IHEP who has worked on the BEPC since its inception. "The project was also very important because it marked the opening of China and the start of collaborations with the U.S. and Europe in basic research." As a result, he added, "the whole country worked together." The BEPC was completed in 1988, with the first electron-positron collisions occurring on October 16 of that year. [66]

Deng and other government leaders inspected the facility later that month, extending their congratulations to the builders. Lee was also pleased by the outcome. "In the success of this project, I see the great confidence of the Chinese nation," he said.[67]

Located on the IHEP campus, next door to the University of the Chinese Academy of Sciences, the original incarnation of the BEPC consisted of an underground, 202-meter-long linear accelerator, which fed electrons and positrons into a "storage ring," 240 meters in circumference. The storage ring was a circular tunnel, lined with magnets, which circulated bunches of electrons and positrons in opposite directions at close to the speed of light and then banged them together in violent crashes at collision energies in the range of 2 to 5 GeV.[68] The beams of electrons and positrons were made to cross each other in the center of the detector, the Beijing Electron Solenoid (BES).

The energy range the BEPC focuses on is sometimes referred to as "charm physics"—the study of particles that contain charm quarks. The field was started, in a sense, by Ting and Richter's 1974 discovery of the J/Psi particle, a meson consisting of a charm quark and charm antiquark. This finding led to a spate of other discoveries, collectively referred as the so-called "November revolution," which confirmed the notion that quarks were real, transforming physics in the process.

But that revolution is not yet played out. "There is [still] a lot going on in that energy region," said University of Hawaii physicist Fred Harris, who has served as co-spokesperson for the Beijing experiment. And in that particular energy regime, Hesheng Chen remarked a decade ago, "we dominate"—a statement that likely still holds true. Some of that dominance, to which Chen refers, may owe to the fact that many of the Chinese physicists leading the effort at the BEPC— including Chen himself and his successor Yifang Wang—were trained in Samuel Ting's lab.

Compared to the 50 GeV collider that had been previously proposed for China, said Chen, the BEPC's "energy was lower but it was more interesting."[69] Electron-positron collisions at the BEPC can be precisely tuned to create charm quarks and antiquarks. The beam energy can be adjusted to generate large numbers of J/Psi particles and excited forms of those particles called psi-prime. Electron-positron collisions in that same energy range, resulting in these particles' mutual annihilation, can also produce the tau lepton—a heavier cousin to the electron and muon.

That's exactly how Stanford physicist Martin Perl discovered the tau in 1975, through collisions of this sort at SLAC. The best estimate of the tau mass at that time—a value of 1784 MeV (or, equivalently, 1.784 GeV)—was made at the Stanford Positron Electron Accelerating Ring (SPEAR) at SLAC, Perl recounted in a personal history of this discovery. "It is only in 1992, fourteen years later, that there has been an improvement in the measurement of m_{tau}; the BES collaboration using the BEPC [electron-positron] collider has just reported [that] m_{tau} = 1776.9±0.5 MeV."[70] The BES team, in other words, had improved the accuracy of the tau mass measurement by a factor of ten.

In 1999, experiments at the BEPC also yielded the most accurate measurements at that time of the "R value" in the 2 to 5 GeV range. An important parameter of the Standard Model, R measures the likelihood of creating hadrons (particles containing quarks), as opposed to creating muons (a kind of lepton), from collisions between electrons and positrons. These measurements, which improved the experimental precision by a factor of three, were made using the BEPC's recently upgraded detector, BESII. Reducing the uncertainty in the R value measurement led to a better estimate of the Higgs boson mass. "Using the old value of R, which had big uncertainties attached to it, the most probable mass for the Higgs boson was below 100 GeV," explains Harris. "When we refined the value of R, we moved the expected Higgs mass value above 100 GeV." He and his colleagues published a couple of papers on this subject around the year 2000. Twelve years later, experiments at the LHC revealed a Higgs mass of 125 GeV, in keeping with the estimates made at the BEPC.[71]

In addition to distinct accomplishments like this, the BEPC has also served a far broader purpose. Thanks to the successful operation of this machine, Chen commented, "a new generation of accelerator engineers and data engineers are growing up [in China].[72]

Indeed, as Panofsky reported, "The success of BEPCI and its detector ... led to plans of further expansion of the high-energy physics program in China."[73] Although more ambitious ideas were considered, it was eventually decided to keep the machine's original tunnel while replacing almost everything else. Two separate storage rings, with two separate rings of magnets, were installed in place of the original single ring. One storage ring was reserved for electrons, and 93 bunches of electrons could fit into it, where just a single bunch fit in before. The other storage ring was reserved for positrons, with 93 bunches of positrons fitting into that ring, too. As a result of packing in so many more electrons and positrons, the revamped accelerator called BEPCII could yield a 100-fold increase in the collision rate, which, in turn, is closely tied to a parameter called luminosity. The essential point is that a great jump in the number of collisions meant that the machine was capable of gathering data at a dramatically higher rate.

The other main push was to enhance the sensitivity of the detector, which was rebuilt once again and subsequently renamed BESIII. The 800-ton BESIII replaced the conventional magnets of its predecessor with stronger superconducting magnets—unlike any fabricated in China before—to measure the energy and momentum of particles found in the aftermath of a collision. These magnets focus the electron and positron beams as they enter the interaction zone in the heart of the detector. The BEPCII also uses superconducting technology to keep electrons and positrons circulating at the proper energy, as all charged particles give off energy as they bend in the presence of a magnetic field—a phenomenon called synchrotron radiation.

Construction of the BEPCII began in January 2004. All the new equipment for the improved collider and detector was designed and built at IHEP, making use of the institute's own machine shop. Construction was finished in 2008, and the first collision events occurred on July 19 of that year. The BEPCII almost immediately set new records for luminosity in the energy range the accelerator operates in. As the *CERN Courier* reported in 2008, following the first collisions, "When fully operational, the BEPCII/BESIII complex will be the world's premier facility for studying properties of charmed mesons and tau leptons." [74]

The hundred-fold jump in luminosity compared to the original BEPC made the BEPCII ten times more luminous than CLEO, a competing machine at Cornell University that shut down in 2008 after almost 30 years of operation. But the upgrade had little to do with bragging rights. Instead, it was all about the science. "With limited statistics, you can find a hint that maybe something is going on," says University of Rochester physicist Edward Thorndike, a BESIII collaborator. "But with ten times more data you can see if it's just a fluctuation or if it's a real effect" —and if the latter is true, "you've found something exciting." [75]

"Thanks to the Chinese, we may be able to answer some of the most compelling questions in particle physics," the University of Minnesota physicist Ronald Poling said after the renovation was completed. "The Chinese have built it, and we have come." [76]

All told, about 350 physicists from some 50 institutions and eleven countries work on experiments at the IHEP facility. [77] The main physics aims for the refurbished complex, according to Hesheng Chen, "are precision measurements of charm physics and the search for new particles and new phenomena, mainly in the energy region of the J/Psi and psi-prime." [78]

In broad terms, the focus of this effort is on testing the Standard Model at low energy. The theory of the strong interaction, quantum chromodynamics (QCD), has been studied much more thoroughly at the high energies that the LHC explores than at low energy, notes BESIII spokesperson Xiaoyan Shen, a physicist based at IHEP. "At the low-energy region we're working in, QCD predicts new kinds of particles. All the particles we know of that are composed of quarks, or hadrons, have two or three quarks. But QCD predicts that you should see hadrons with four, five, or six quarks, or hybrid states with two quarks and one gluon, or maybe just gluons"—a composite particle of two or more gluons bound together, which is sometimes called a glueball. The existence of glueballs is an important prediction of the Standard Model and QCD that has not yet been experimentally verified. Finding them would pose a very good test of QCD theory, says Shen. "And right now, this [the BEPCII] is the best place for finding these kinds of states and particles."[79]

Big news was reported on June 17, 2013, when members of the BESIII collaboration and the BELLE collaboration at the High Energy Accelerator Research Organization (KEK) in Tsukuba, Japan independently claimed evidence for a particle called $Z_c(3900)$ composed of four quarks. Both groups made these particles by accelerating electrons and positrons almost to the speed of light and then smashing them into each other. The BESIII team had an advantage in this regard because it could tune its machine (BEPCII) exactly to 4.260 GeV in order to produce another particle called Y(4620), which consists of a charm quark, a charm antiquark, and a gluon. The BELLE team produced Y(4620) particles as well, but could only do so through indirect means because their accelerator (KEKB) operates at higher energies.

Y(4620) had been discovered in 2005 at the BaBar experiment at SLAC, but the BESIII and BELLE researchers saw something new: Y(4620) lives only about 10^{-23} seconds, and among its decay products they came across a charged particle that had never been spotted before, $Z_c(3900)$.[80] Additional corroboration of this discovery came from the analysis of data from the now-defunct CLEO accelerator.

The fact that $Z_c(3900)$ contains four quarks is not in dispute, but there is no agreement so far as to what $Z_c(3900)$ really is. It could be an actual four-quark ("tetraquark") particle or it could be a subatomic version of a molecule—the conjoining of two mesons, each containing a quark and antiquark. Ahmed Ali, a physicist at the DESY high-energy physics laboratory in Germany, favors the tetraquark picture because a "molecule" composed of two mesons should occasionally split into two halves, which has not been seen.[81]

For now, the correct interpretation remains up for grabs. But no matter which way the verdict swings, $Z_c(3900)$ would still be "a new form of matter," maintains University of Pittsburgh physicist Eric Swanson. And figuring out its true nature should tell physicists a lot about the various ways in which the strong force binds quarks.[82]

In the meantime, BESIII investigators will continue to pore over their existing data, while also generating new data, in the hopes of unmasking the genuine character of $Z_c(3900)$. They will also study other four-quark candidates, including the electrically-charged $Z_c(4020)$—which was recently discovered at BEPCII using a method similar to the approach that uncovered $Z_c(3900)$—and the electrically-neutral X(3872), which was discovered by the BELLE experiment more than a decade ago. Starting from decays of Y(4260), a family of four-quark objects has begun to appear, says Fred Harris. "While the theoretical picture remains to be finalized, more and more clues are suggesting that we are witnessing new forms of matter. And while a new 'zoo' of mysterious particles is emerging, it seems a new classification system may soon be at hand to understand it."[83]

It's not just a matter of adding more and more particles, notes Ryan Mitchell, a BESIII collaborator from Indiana University. "We're starting to see patterns." [84]

Filling out this picture, which still has many puzzling aspects, should keep BEPCII investigators busy for many years. The accelerator is scheduled to run until the early-to-mid-2020s, says Xiaoyan Shen. "We want to keep these physicists occupied, doing vital experiments, while we prepare for the next generation, high-energy machine." [85]

Harris, for one, expects to keep coming to Beijing for a while. "There are some unique things we can do at BEPCII, he says, such as tuning the machine to crank out Y4260 particles in vast quantities. "As long as the Chinese government will support it, there is a lot of interesting physics to be done here." [86]

The question looming above everything else pertains to what will replace the Beijing facility, which has been operating, in one form or another, since 1989. A related question is whether China is ready to move forward with a much bigger collider, operating first as an electron-positron accelerator and later as a proton-smashing machine. Most of the key Chinese physicists pushing for a new collider have previously worked at major facilities, like CERN'S LHC or Large Electron-Positron Collider (LEP), and perhaps at the more modest sized BEPC as well. [87]

Robbert Dijkgraaf, director of the Institute for Advanced Study, thinks that China is well positioned to take the initiative in this area, and he's "very enthusiastic about the prospect of China stepping up." At the moment, he says, "It has the unique ability as a single country to actually support very large-scale infrastructure programs." Part of this readiness stems from the deeply held conviction "that science is the way to make progress," Dijkgraaf adds. "China may be the only country that takes a long-term view of science, looking 40 to 50 years ahead," which is essential in high-energy physics, where it can take more than a single generation to find answers to pressing questions—the 48-year-long quest to find the Higgs boson being a case in point. [88]

IHEP's Chuang Zhang, who has been designing colliders in China for more than 40 years, also believes that the time has come for his country to start thinking big. "This is part of our long-term dream, going back to Project 753, when we wanted to catch up to the world's leading labs like CERN and Fermilab at the high-energy frontier," he says. "We must have the dream in order for it to come true. But we have to work hard in order to have the chance of realizing that dream." [89]

Harris, who has pursued particle physics in Beijing over the past 20 or so years, is not prepared to dismiss that dream out of hand. Building a machine on the order of the so-called "Great Collider" would pose immense challenges, technical and otherwise, he acknowledges. And China would definitely need outside help to pull off such a feat. On the other hand, Harris says, "I was worried about BEPCII and BESIII, but China was able to complete those upgrades on budget

The linear injector of the Beijing Electron Positron Collider (Photograph courtesy of IHEP)

The control room of the Beijing Spectrometer or BES (Photograph courtesy of IHEP)

Samuel Ting inspecting equipment at the Beijing Spectrometer (Photograph courtesy of IHEP)

A view of the Beijing Spectrometer III or BES III (Photograph courtesy of IHEP)

and on time, and they were able to build superconducting magnets on their own."

Going from there to the Circular Electron-Positron Collider, and to the Super Proton-Proton Collider after that, would be a much bigger step—a quantum leap, if you will. Of course, China has not stayed still in the three-plus decades since the BEPC was conceived and built. "There has been tremendous growth and improvement in that time," Harris says, "not only in physics but throughout the whole country, up to the point that the U.S. Department of Energy is starting to take the possibility of a Chinese collider seriously." And based on his considerable experience in this matter, and lengthy involvement with China's particle physics program, Harris is taking that possibility seriously as well. [90]

Chapter 5

An Accelerator for—and of—the World

WHEN THE SUPERCONDUCTING SUPER COLLIDER (SSC) WAS CANCELLED IN 1993 all was not lost, for plans for the Large Hadron Collider (LHC) in Geneva were still moving forward. But some physicists and policy makers began to wonder, out loud, whether this machine, soon to become the most powerful particle accelerator ever built, would be the last of its kind—the "Last Hadron Collider," as it was sometimes called. Part of that speculation had to do with the sizeable price tag attached to this mammoth facility, with its cathedral-sized detectors and 27-kilometer-long ring carved out of rock, 100 or so meters below the surface of suburban and exurban Geneva. Approximately $8 billion dollars had been spent getting the collider ready up to the time that the first proton beams started circulating in September 2008.[1] The other part of the discussion revolved around a physics question—namely, would there be enough vital science left to do after the LHC closed up shop in 2035, or thereabouts, to justify the considerable expense of an even bigger, costlier successor?

In 2012, three years after the first proton collision occurred at the LHC, scientists in Geneva announced the discovery of the long-sought Higgs boson—a finding that was said to have put the finishing touches on the Standard Model. For a while, people were content to revel in this breakthrough, both by celebrating the marvelous achievement as well as by pondering the many implications and questions that it raised. But within a matter of weeks, some physicists had shifted their gaze towards the future. The LHC was slated to run for at least two decades more, and a cadre of researchers was already thinking about what machine, or preferably machines, would come next. They refused to accept the notion that the LHC represented the end of the line—four decades after the world's first hadron collider, CERN's Intersecting Storage Rings (ISR), had begun smashing proton beams into each other. To them, the Higgs discovery was a beginning—an accomplishment that opened the door to a new chapter in physics—rather than an ending.

Looking ahead, these physicists felt there was still pressing work to be done on at least two broad fronts: They wanted to study the Higgs boson, which was unlike any particle we'd ever seen before, in far greater detail, gaining much sharper views than the LHC could afford. And they also wanted to explore the high-energy frontier in the 14 to 100 TeV realm, which was beyond the reach of the LHC in its present form and therefore stood as a kind of terra incognita.

The researchers were encouraged by the fact that the Higgs mass was relatively low, on the order of 125 GeV, which meant that an electron-positron accelerator (with an energy much lower than the LHC) would be ideal for producing Higgs bosons in vast quantities without having to contend with the messy background that inevitably comes from slamming protons—complex mixtures of quarks and gluons—together. The Higgs particle's low mass was a plus in this regard because it meant that a collider, which yielded center-of-mass collision energies of about 240 GeV, would be more than adequate to serve as a "Higgs factory."

It also meant that a circular collider could be a practical choice. Losses from synchrotron radiation—the rate at which energy (in the form of X-rays) is given off by "relativistic" charged particles, traveling in a circle at close to the speed of light—scale with the fourth power of the beam energy. Synchrotron radiation losses at a 240 GeV circular collider would not be unmanageable, whereas they could be overwhelming at 500 GeV and up, or at least very difficult and costly (in terms of exorbitant power demands) to deal with.

The rate of synchrotron radiation losses is also inversely proportional to the square of the circumference of the circle the particles are constrained to travel around. So there could be some advantages in building a large-circumference ring—perhaps on the order of 100 kilometers as opposed to, say, 27 kilometers. A tunnel that big would afford an additional benefit: It could later be used for accelerating protons, yielding collisions on the order of 100 TeV, or about seven times more powerful than the LHC. (Synchrotron radiation is a much less significant factor for a hadron collider because radiation losses are inversely proportional to the fourth power of a particle's mass. Given that electrons and positrons are about 2,000 times lighter than protons, they lose energy about ten trillion times faster than protons.[2])

In January 2012, Chinese physicists discussed the possibility of building a circular electron-positron collider that would serve as a Higgs factory—an idea that ran contrary to conventional wisdom at the time. "The next 'big' collider after the LHC has been the subject of the international HEP [high-energy physics] community for more than a decade," wrote University of Geneva physicist Alain Blondel and his colleagues in 2013. "Since then ... under the direction of ICFA [the International Committee for Future Accelerators], a series of decisions have been made"—namely that the device in question "should be a *linear* [electron-positron] collider."[3] The Chinese plan challenged that assumption, arguing that a Higgs factory was well within the reach of a circular collider. Blondel and others had, in fact, promoted this same concept before—and revived it in August 2012 after the Higgs discovery—suggesting that a new machine called LEP3, optimized for producing Higgs particles, could be installed inside the LHC tunnel (the site of the original Large Electron-Positron collider or LEP).

During a September 2012 meeting, the Chinese team decided that building a Higgs factory alone might not yield enough scientific return to justify the considerable investment involved. At the suggestion of IHEP head Yifang Wang, they decided to add another element to their plan in order to "sweeten the deal." Their first step would be to begin, as before, with the construction of a Circular Electron-Positron Collider (CEPC) whose main purpose would be to study the Higgs. The next big component would come many years down the road when a second machine, a Super Proton-Proton Collider (SPPC), would be installed inside the same tunnel. The beauty of this plan is that the Chinese researchers would start with the electron-positron technology with which they were most experienced, while allowing a reasonable amount of time to develop the far less familiar technology that would be needed for a proton collider running at much higher energies. The overwhelming consensus reached at the meeting was that this proposal warranted further study.

At a Fermilab conference two months later, Qing Qin, the director of IHEP's accelerator division, introduced the idea of a two-phase accelerator—a circular Higgs factory and proton-proton collider, which shared the same tunnel. The response was quite enthusiastic, says Wang. "The idea immediately caught the attention of almost everyone."[4] He decided to formalize the program in the next year, arranging for an official study group—made up almost exclusively of Chinese scientists—to look into this possibility in a more rigorous fashion. More than 120 physicists from 19 Chinese universities and institutes participated in the study. A few months later, the Center for Future High Energy Physics was founded in Beijing, under the direction of Nima Arkani-Hamed, to help build the physics case for a large circular collider—an exercise deemed worthwhile regardless of where on the planet a device of this kind might ultimately be situated.

While there was a clear preference, of course, to see this facility take shape in China, it should be stressed that the CEPC and SPPC were never intended to be "Chinese colliders" *per se*. Instead, the plan from the outset was for China to be the host of an international facility, which physicists from all over the world would use, working alongside their Chinese counterparts.

At roughly the same time, a Future Circular Collider (FCC) study was initiated at CERN, charged with developing an "ambitious post-LHC accelerator project."[5] The CERN physicists were also contemplating the idea of an entirely new facility that would operate in two distinct stages: an electron-positron machine, followed by one for accelerating protons. As an alternative, they considered forgoing the first stage and building a powerful proton accelerator right from the start. (This proposition was originally referred to as "TLEP," the Triple-Large Electron-Positron Collider, for the tunnel being considered would be roughly three times as long as the LHC—and former LEP—tunnel.) If any of these projects came to fruition, the LHC would not be the "Last Hadron Collider," and the future of high-energy physics would be assured for decades to come.

The prospect of a large Chinese collider excited Xinchou Lou, a physics professor at the University of Texas at Dallas who originally came to the United States from China to pursue graduate and postdoctoral studies. The university's physics department offered Lou a job in 1993 to work on the SSC, but by the time he joined the faculty in January 1994, the Super Collider project was dead. His only hands-on experience with the SSC consisted of removing some equipment from its abandoned laboratories in Waxahachie and bringing them back to Dallas. Lou has held a joint appointment at IHEP since 2012, when he came at the request of Yifang Wang to become director of the CEPC-SPPC project. Lou took the job in the hopes of making the Chinese venture a success, while avoiding some of the pitfalls—cost overruns and other administrative problems—that had contributed to the SSC's demise. He felt like this project "had a real shot," given the dedicated group of Chinese physicists he had met "who were committed to doing real science," and given the supportive response he sensed from funding agencies and from the government at large.[6]

Here is the general plan that Lou, Wang, and their colleagues have come up with so far, although the design is still somewhat malleable: While the initial proposal calls for a tunnel at least 50 kilometers in circumference, the real hope to is build a 100-kilometer long tunnel, assuming that sufficient funds are available. The tunnel would be 50 to 100 meters deep and about 6 meters wide—wide enough to accommodate the CEPC, SPPC, and a separate "booster" accelerator mounted to the ceiling. The purpose of the booster is to accelerate protons to an energy at which they could be injected into the high-energy ring.

The plan is to keep the CEPC in the tunnel after the SPPC is built and operating. At times, the two accelerators could be run simultaneously, making collisions between protons (or heavier particles like gold ions) and electrons and positrons possible —in order to investigate, for example, unusual states of matter called quark-gluon plasmas, thought to have existed during the universe's first few microseconds.

The giant tunnel in which these particles are accelerated, like that of the LHC, would be round but not an exact circle. Instead, it would consist of eight arcs (comprising about 85 percent of the circumference) and eight shorter straight sections (comprising about 15 percent of the circumference). Four of the "straights" will include the interaction zones, each with its own detector—two reserved for the CEPC and two for the SPPC. The other four straights are places where particles are accelerated and injected into the main ring. The curved sections are mostly filled with magnets needed to bend the beams of charged particles. These magnets are not needed in the straight segments (where no bending is required), which would leave room for the detectors and other key equipment.

A 240 GeV electron-positron machine (the CEPC) would be installed first. After about five years of research and development, according to the current timetable, construction of the CEPC could start as early as 2021, with operations beginning in 2028. Meanwhile, research and development on the SPPC would be carried out from 2015 to 2030. A key focus of this work would be devoted to readying high-strength superconducting magnets. Work on the engineering design would then proceed from 2030 to 2035, followed by construction from 2035 to 2042. If the schedule holds, data collection could begin in 2042.

That's a long way off—which is good, because China needs time to prepare for its first-ever hadron collider. Starting with the CEPC makes a lot of sense because electron-positron colliders are cheaper to build, and are also more in line with the expertise that has been accumulating in China since the Beijing Electron Positron Collider was turned on in 1988.

"Circular and linear colliders are complementary ... with different pros and cons," says CERN physicist Fabiola Gianotti.[7] That suggests that both the CEPC and the International Linear Collider (ILC) could have important contributions to make to high-energy physics.

A circular collider like the CEPC has the potential to create roughly ten times more Higgs bosons than a linear collider creates, and thereby generate much more data.[8] On the other hand, the ILC, which is initially slated to operate at 500 GeV and may later be upgraded to 1 TeV, "has the advantage of being able to go to higher energy" and therefore to seek more particles beyond the Higgs, says IHEP physicist Jie Gao. A linear collider, moreover, can be extended to reach higher energies, whereas a circular collider cannot be expanded, once in place, without building a whole new tunnel. One drawback of a linear collider is that it has just one interaction point, adds Gao, "so you normally have room for just one detector. Usually in accelerators we like to have two or more detectors to do cross checks." The LEP had four interaction points, as does the LHC, and the plan for the CEPC includes four interaction points as well. The ILC could have more than one detector, Gao notes. "But the detectors would have to be switched at the ILC, which is less convenient." [9]

As for the physics that might come out of the CEPC and eventually the SPPC, we won't go over all of the ground covered in Chapter 3 of this book, but we will revisit a few points and touch on others not stressed before. The science case for a new collider, which might eventually reach 100 TeV, may not be any less compelling than the case made for the LHC prior to its construction, but the argument is definitely more subtle. The LHC was sometimes called the "no-lose accelerator" because at the power at which it was designed to operate—initially 8 TeV and eventually increasing to 13 or 14 TeV—many physicists were confident that it would either see the Higgs boson or discover whatever else was responsible for giving elementary particles their mass. [10] This conviction was based, in part, on the so-called "no-lose theorem," which maintained that the accelerator was practically the closest thing to a sure bet you could find in physics.

The gist of this idea went as follows: Proton collisions at the LHC will sometimes produce a pair of W bosons, and these W bosons will occasionally collide with each other. In a theory without a Higgs boson, the probability that two W bosons will collide goes haywire: it keeps going up as the ambient energy increases, eventually becoming bigger than one. An interaction with a probability greater than one is a nonsensical situation in physics—and in the world at large. It simply cannot happen. When things like this start showing up in your calculations, it's a sign that your theory is in trouble or is, at the very least, in need of repair.

A Higgs boson, however, could save the day by keeping the W-W scattering probability to a relatively low value, well below one. But the Higgs could only ensure this happy outcome, calculations have shown, if its mass is less than 800 GeV. Given that the LHC was designed to look for particles with masses up to about 1000 GeV (or 1 TeV), the collider was almost ensured of making a new discovery—either seeing the long-sought Higgs particle or finding something even more extraordinary. However things shook out, the LHC seemed destined to uncover some interesting physics. [11]

"This no-lose scenario does not exist for 100 TeV," says CERN director-general Rolf Heuer. [12] Nevertheless, he believes there's still a strong argument to be made for such a machine, even if it can't be summed up in just three words—finding the Higgs—which was the case for its predecessor.

For even though physicists have finally gotten their hands on the Higgs boson, they still don't understand what they've really got. The Higgs discovery in 2012 meant that the examination of this particle could begin in earnest, along with a chance to explore the many and interlocking mysteries tied up in it. One question of more than passing interest has to do with why the Higgs field is "spreading through all space, touching every particle, and giving it mass," says SLAC physicist Michael Peskin. Perhaps the best way of getting more clues about this field and related puzzles is to study the Higgs boson with exacting precision. [13]

That will, in fact, be a central mission of the proposed CEPC, which is expected to generate millions of Higgs particles, enabling physicists to observe with unparalleled clarity their interactions with other particles and with themselves, and to witness rare decay processes that the LHC lacks the sensitivity to see. Better statistics will also yield a sharper measurement of the Higgs mass, which can, in turn, lead to more accurate predictions for various decay modes and production rates. Researchers would then look for any signs of deviation from the precepts of the Standard Model. By pooling the data acquired in this way, they could construct a fuller picture of the Higgs, which should help them understand whether this boson is truly elementary, whether it might harbor some hidden, microscopic structure, or whether new particles or forces might come into play in the course of its myriad possible decays.

SPPC investigators will pursue some of these objectives, while also tacking some additional items to their list. First and foremost, the new proton collider will put them at the forefront of the hunt for new particles that may have been predicted by theories involving supersymmetry and dark matter—or whose existence, unrelated to those ideas, catches us completely off guard. The researchers will have a chance to learn more about the imbalance between matter and antimatter, and how it initially arose, as well as gaining insight into the phase transition responsible for electroweak symmetry breaking. All of these ideas are critically important to our conception of the universe. Yet they are also closely intertwined and might even be thought of as different aspects of the same question.

The SPPC's five- to seven-fold increase in center-of-mass energy relative to the LHC would enable scientists to search for fundamental particles that are, accordingly, some five to seven times heavier than those that can be produced by the latter machine (unless, of course, further upgrades are made there). [14] Particles that the LHC can turn out in relatively small quantities might be produced at up to a thousand times higher rate at the SPPC, resulting in a 100-fold jump in resolution. At least some of these new particles, assuming any are found, might be connected to the notion of supersymmetry. Supersymmetry models commonly predict that a 100 TeV collider should be able to identify the "superpartner" of the top quark, also known as the "top squark" or the "stop," extending the search for that particle up to about 8 TeV. [15]

One difficulty that lies in the way of finding the stop is that the stop's experimental signature is quite similar to that of the top quark. Ashutosh Kotwal, a Duke physics professor now on assignment at Fermilab, finds it ironic, as well as somewhat amusing, that the top quark, "the same object that created so much happiness [when it was discovered at the Tevatron in 1995] is now the anonymous background from which we have to pick out its supersymmetry partner." That is one way of measuring progress in this research, says Kotwal, the U.S. coordinator of a global effort to build a large new proton collider. "Yesterday's discovery soon becomes the background for the next discovery."

The general search strategy is based on the notion that some top quarks are produced from the decay of their heavier (and still hypothetical) superpartner, the stop, and that some are produced through other mechanisms provided by the Standard Model. "Because you're seeing top quarks coming from both of these pathways, it becomes a difficult job," Kotwal adds, especially because the LHC already produces billions of top quarks and a 100 TeV machine would turn out even more. But among that crowd of particles, he says, "a 100 TeV machine should also produce huge numbers of stops, and we would be able to study them in great depth." [16]

The discovery of the supersymmetric version of the top quark would open the floodgates to additional discoveries, as it would provide a clear indication that there are more than a dozen other supersymmetric particles out there, waiting to be found. Among them, however, should be a special treasure: For the lightest supersymmetric particle, in principle, cannot decay down to anything else and therefore should be perfectly stable. If the whole notion of supersymmetry is correct, this particle may constitute the bulk of the universe's dark matter. And the general assumption is that it would come in the form of a weakly interacting massive particle or WIMP.

The SPPC would facilitate the search for dark matter via some of the approaches discussed in Chapter 3. We could, for instance, attempt to produce squarks (the superpartners of quarks) or gluinos (the superpartners of gluons), which would, according to theory, keep decaying down to lighter and lighter supersymmetric particles until dark matter might eventually be spotted in the form of a "missing energy" signal during a collision event. But a proton collider like the SPPC might also produce dark matter directly in the form of particle pairs. Supersymmetry could then suggest candidate particles that physicists might look for. Among the leading candidates are winos, the postulated superpartners of the W bosons, and Higgsinos, the superpartners of the Higgs boson.

This approach may offer the most promising way of spotting winos and Higgsinos because these particles may be difficult to capture in "direct detection" experiments employing underground tanks of liquid xenon. [17] And even if a subterranean detector could mark the passage of an errant wino or Higgsino, it probably could not say much about that particle's properties. That's one way in which a high-energy accelerator offers distinct advantages over other dark matter search strategies. "What a collider can do is make these particles, reproducing what must have happened at the beginning of the universe," says Kotwal. "And if these particles come from the decay of other particles, you are making them too. Which means that the scope of what you can learn at a collider is much, much more." [18] Winos and Higgsinos, furthermore, are likely to have masses around 1 TeV, which could put them out of the LHC's reach but well within the grasp of the SPPC, giving it a much better shot at success. [19]

The machine will also go deep into the territory where other dark matter candidates are expected to be found. While there are few guarantees in this business, says Nathan Seiberg of the Institute for Advanced Study, "there is strong evidence for dark matter and strong evidence that it would be found in the energy range [up to about 100 TeV] that we hope to explore here." [20] Of course, producing dark matter candidates in an accelerator would merely be the first step, adds Peskin. "To prove that a particle observed at a collider really is the dark matter particle [that fills the universe], we will need to observe it in our galaxy or as a particle that hits Earth from space." [21]

The SPPC may also have important things to say about another pressing mystery—the matter-antimatter asymmetry and where it came from. Based on the Standard Model, the universe should not be matter-dominated. There should be equal parts of matter and antimatter, which would eventually meet and annihilate each other, emptying out the universe and everything in it. But we are here, standing on terra firma and gazing out at stars and galaxies that extend as far as our eyes (and our telescopes) can see. Yet, in the entire expanse that we've surveyed, there appears to be little if any antimatter around, which raises the question: What happened to it?

In 1967, the Soviet physicist Andrei Sakharov proposed a theory to explain why "the universe is asymmetrical with respect to the number of particles and antiparticles." [22] Sakharov, an eventual Nobel Peace Prize winner, laid out three conditions that, if satisfied, could allow for a universe starting out with the same amount of matter and antimatter to evolve to one in which matter dominates. The first condition is that there must be processes in nature that can change the "baryon number." The baryon number is defined as the number of baryons (like protons and neutrons) minus the number of antibaryons (like antiprotons and neutrons). Given that baryons are bound states of three quarks and antibaryons are bound states of three antiquarks, the baryon number can also be defined as the quark number (the number of quarks minus the number of antiquarks) divided by three. This number is thus +1 for individual baryons (consisting of three quarks), -1 for antibaryons (consisting of three antiquarks), and 0 for mesons (consisting of a quark and antiquark).

If the universe began with the same number of quarks and antiquarks, the baryon number for the whole universe must have been zero to start with. But if matter now dominates the universe, as appears to be the case, the overall baryon number must be nonzero and positive. That could happen only if processes that can change the baryon number occur in nature—processes, that is, in which the baryon number is not conserved. The baryon number could change, for example, if protons and antiprotons were not entirely stable. Some supersymmetry and so-called grand unified theories predict that protons do in fact decay, but this phenomenon has never been observed in the experiments conducted to date. (Maybe we just have to wait a bit longer, or maybe there's a problem with those theories.)

Sakharov's second condition holds that there must be naturally occurring reactions, so-called CP violations, which treat matter and antimatter differently. Some experiments—including those involving neutral kaons (see Chapter 4)—have already shown this statement to be true. In 2004, the BaBar collaboration at SLAC and the Belle collaboration at the KEK laboratory in Japan observed direct CP violations in the decay of neutral B mesons, which are particles composed of a bottom antiquark and either a down or strange quark. The asymmetry, as measured by the BaBar team, was 100,000 times stronger in B mesons than in kaons. [23] These findings have led theorists to believe that CP violations are much more prevalent than has been witnessed to date, and are likely to accompany the weak decays of other particles as well. Other experiments, such as those focused on neutrinos, are actively exploring this possibility.

Sakharov's third and last condition states that the processes that result in a baryon number violation must occur when the universe is out of thermodynamic equilibrium. Physicists have since suggested that the original arrangement, with matter and antimatter equally abundant, could have been disrupted during the phase transition through which the Higgs field acquired a nonzero value everywhere, and electroweak symmetry was simultaneously broken. One baryon number-violating process that might have taken place during the electroweak phase transition, according to theorists, involves squeezing two protons together so hard that they turn into an antiproton and some leptons instead. [24] Ever since that transition, things would have been irrevocably altered: matter would have ruled the universe, the previously unified electroweak force would have split into two separate forces, and the once interchangeable carriers of the electromagnetic and weak forces would have become distinct.

In order for this scenario to be true and fulfill the Sakharov requirements, the electroweak phase change must have been of "first order." A first-order phase transition is, like the boiling of water, an abrupt and dramatic change of state. It is an out-of-equilibrium process that occurs in one direction and is not readily reversed. The fact that it is irreversible is essential to Sakharov's argument, for otherwise the matter-antimatter asymmetry might have been erased almost as soon as it was created rather than being frozen in place.

Many strands of evidence have pointed physicists towards the electroweak phase transition as being the moment in cosmic history in which this disparity arose (although there are many other hypotheses that have appealing features as well). Maybe the balance between matter and antimatter was upset during this brief interval when the universe, itself, was out of equilibrium and hence out of balance too. If matter came out of that transition with just a slight edge over antimatter, that could have been enough to wipe out virtually all traces of antimatter, which is pretty much what is seen today. The only problem with that scenario is that the measured value of the Higgs boson mass—about 125 GeV—is not consistent, according to the Standard Model, with a first-order phase transition. Instead, the Standard Model tells us that a Higgs with that mass would lead to a much smoother, more continuous "second-order" phase transition.

But all is not lost, says Kotwal. "To get a first-order phase transition, which in turn can lead us to the matter-dominated universe we see, we just need to change the Higgs a little bit from the picture that the Standard Model provides." Precise measurements are thus needed to detect even slight modifications in the Higgs properties—modifications that might be enough to have tipped the balance in matter's favor. [25]

One of the best ways of seeing possible departures from predicted behavior would be to study the way Higgs particles interact with one another. Three-way interactions—in which one Higgs particle comes in and two come out—may be particularly revealing. The three-Higgs coupling is intriguing, says Peskin, "because Higgs models that predict a first-order phase transition also typically predict a large shift of the three-Higgs coupling." However, he adds, this coupling is much harder to measure than other Higgs couplings. [26]

The Standard Model predicts how strongly the Higgs particle should interact with itself. There is only one Higgs particle in that model, but if we modify the theory to enable more than one kind of Higgs particle to interact—which supersymmetry, among other ideas, allows—then the observed rate of self-interaction should change. If there is a second Higgs particle, and maybe more, a 100 TeV collider should be able to make it, Kotwal notes. "And we know that it is rather easy to construct a modified theory, with two or more Higgs, which allows for the electroweak phase transition to be first-order. That, in turn, could tell us why we're here and why antimatter didn't wipe out everything."

So the issues at stake for a 100 TeV machine couldn't be much higher. We want to understand the Higgs particle and the associated Higgs field, without which atoms and molecules—and every object of substance made up of them—would never have taken shape. We also have a shot at discovering the identity of dark matter, without which—most theories hold—stars and galaxies would never have materialized, nor would have we. Our existence also depends on the universe being matter-dominated, even though current theories can't yet explain why this is the case. That mystery, too, may soon be resolved. "These are incredibly important questions, and answering them will result in a quantum increase in our existing knowledge," Kotwal affirms. "Once you see all that, you're likely to ask: How fast can we get there? How fast can we get these new colliders which can help us explain really big mysteries that relate to the whole universe and why we're here?" [27]

The answer—at least in terms of the CEPC and SPPC—is that we will, unfortunately, have a bit of a wait, given that this project is still just an idea rather than anything concrete. We ought to push forward aggressively, while at the same time proceeding in a thorough and systematic fashion, as good science—and especially science at the high-energy frontier—cannot be rushed.

As to the question of where the Chinese collider stands at the moment: A Preliminary Conceptual Design Report, drawing on contributions from some 300 scientists from 57 institutions in nine countries—was completed in March 2015, having been inspected by an International Review Committee prior to its release. The three-volume report, totaling more than 600 pages, discussed technical details concerning the accelerators and detectors, as well as determining the highest-priority physics questions that the CEPC and SPPC should address. Engineering issues are also taken up. For example, says Xinchou Lou, "We might look into what is needed, in the way of logistics, to build the tunnel. What are the power and ventilation requirements?" He and his fellow researcher need to be able to answer questions like this, as well as identify critical technical questions that must be answered during the research phase. "So this is a very important process for us to go through in order to do to convince ourselves and the community and the government that we can carry this project forward," Lou says. [28]

So far, the international review has been positive. "The pre-CDR was great—a blueprint of where fundamental physics stands today and what this machine could accomplish," says David Gross, a Nobel laureate in physics who has been to China dozens of times to meet with leading scientists and government officials. "There's been an incredible amount of work done already for a project that hasn't been approved yet or received research and development funding. I'm really hopeful, and it will be important to get something on the table soon." [29]

Members of the CEPC-SPPC team submitted a proposal to the Chinese government on April 15, 2015, seeking five years of research and development support, and they are now waiting for a response. The five-year period will not only allow time for critical details to be worked out, it will also provide an opportunity to see what turns up during the second run of the LHC, which resumed high-energy collisions in June 2015, quickly reaching the 13 TeV mark.

Knowledge of what the LHC sees, or doesn't see, will have a powerful influence on the research agenda at China's collider. If new particles are discovered in Geneva, a 100 TeV machine could produce them in even greater quantities. And there are likely to be other particles out there, lurking somewhere beyond the LHC's reach. On the other hand, if no particles are found during Run 2, there will be greater urgency in the exploration of the 14 to 100 TeV range, in the hope that some light may be shed on the mysteries brought to the fore by the Higgs discovery.

Meanwhile, the proposed China collider is not the only idea being pursued at the moment. In September 2013, a five-volume technical design report for the ILC was completed, presenting details about the design of the accelerator and detectors, their anticipated performance, and the cost of such a facility. A subsequent report was then undertaken for a specific site, the Kitakami Highlands region of northern Japan.[30] "The design has been done, and they're ready to put a spade in the ground," says Peskin, an avid proponent of the project. Although the ILC is far ahead of other future accelerator concepts, it has not yet received full government backing or financial backing. And achieving a national consensus in Japan, Peskin notes, "is turning out to be a very slow process."[31]

CERN, meanwhile, is formulating its own plans for an electron-positron collider. The Compact Linear Collider, or CLIC, would be 50 kilometers long and employ novel acceleration techniques to achieve collision energies of 3 TeV. Although CLIC would be bigger and more powerful than the ILC—which is initially supposed to operate at 500 GeV, with an eventual upgrade possibly boosting that to 1 TeV—CLIC is not nearly as far along, having only reached the conceptual design phase.

The aptly named Very Large Hadron Collider (VLHC) was originally proposed as a 175 TeV collider that would accelerate protons in a circular tunnel, 233 kilometers around, located near Fermilab in northern Illinois. A 272-page formal "Design Study" was completed in June 2001,[32] and the concept has been discussed intermittently since a meeting at Snowmass, Colorado held a month later during which Peter Limon, the former head of the technical division at Fermilab, gave a presentation about it.[33] Although the VLHC has not taken root in the intervening years, it has not disappeared entirely either. Before a U.S. government advisory panel in November 2013, Peskin discussed a significantly downsized version of that idea—a 100 TeV collider with an 80-to 100-kilometer tunnel—in an attempt to sketch out the physics case for such a machine in the post-Higgs-discovery era.[34] Unfortunately, the VLHC is not under serious consideration at the moment as a U.S. venture —perhaps a reflection of the dormant state of affairs in the nation's high-energy particle physics program.

The Future Circular Collider (FCC) study, on the other hand, is a much more credible proposition than the VLHC, having the backing of CERN—the organization that runs the world's largest and most productive particle physics laboratory. Like China's plan, it concerns a circular collider centered around Geneva, with a circumference of 80 to 100 kilometers, which would be capable of generating collision energies up to 100 TeV. The FCC team is expected to deliver a conceptual design report to the European Commission for Future Accelerators by 2018.

University of Pisa physicist Guido Tonelli, a member of the CMS team at the LHC, believes it is important to move forward with the new collider soon, regardless of what might be found at the LHC in the next few years. "If nothing appears in the next phase of the LHC, we have to move to higher energies, because there we might find solutions to the big questions that are still open," he says. "If we do find something, we know that at the LHC we might [only] be able to see the 'tail of the dinosaur,' and we would need a machine with much higher energy to see the entire animal. We don't yet know the details of the next accelerator, but the need for one is clear." [35]

Rather than posing a threat to China's proposed collider, a credible effort like the FCC may, in fact, benefit the Chinese project. "Having other ideas out there is helpful," says Yifang Wang. "It shows our government that there is enormous interest in a project of this sort. Without competition, there is no urgency to move forward." Stasis could set in, which would be detrimental to the field. [36]

Competition is healthy, Nathan Seiberg agrees. "From the physics perspective, it's essential that there is a machine, and it's much better if there are two machines rather than one. Historically, we've seen that progress is much more rapid and results are much more reliable when there are two independent machines." As an example, he cites the discovery of the J/Psi particle by Samuel Ting and Burton Richter, at two different machines, using two different technologies. "The fact that they saw the same signal left no doubt." [37]

There is some doubt, however, as to whether there is enough money and technically knowledgeable people in the world to construct and operate two 100 TeV proton machines at the same time. Arkani-Hamed would like to see more than one new collider built, though he hopes that Europe and China won't duplicate each other's efforts. "The best thing that could happen for the field would be for every group of people, and every region of the world, to try to do the most ambitious thing they can to move science forward. And this is by far the most ambitious thing that China can do—less so perhaps for Europe and the United States because they have built big machines of this type before." [38]

But all of the aforementioned projects pose challenges. The ILC, by most accounts, is ready to move forward—and has been so for a while—but political resolve is still lacking. Although CERN would clearly be up to the task of building a 100 TeV machine, if any organization could, it will have money and manpower tied up in the LHC for the next 20 years, possibly putting a crimp in plans to take on a massive new initiative. The potential site for a 100-kilometer tunnel "is not ideal," according to CERN's former director-general Luciano Maiani, "with the mountains on one side and the lake on the other, but maybe it can be done." [39] China faces its own challenges of a different nature, but equally if not more formidable. Simply put, it has never undertaken a large-scale accelerator project before, nor does it have any experience with high-energy proton colliders. Lastly and sadly, the VLHC, as a U.S.-led proposition, appears to be an extremely remote prospect, given the county's plans to put neutrino studies at the center of its domestic particle physics program (apart from the more than 1,700 U.S. physicists currently engaged in research at the LHC[40]).

In summing up the situation, Peskin considers it "very unclear as to whether any of these machines will happen." The main problem, he says, is not technical feasibility but rather whether any government is willing to put up the money necessary to make such a large project happen. "And if you're not sure what road will get you there, you have to go at least part-way down all the roads."[41]

Historical precedent amply supports that advice. CERN was committed to the LHC in the late 1980s, even though construction was imminent on the much bigger SSC, which some thought could have made the LHC practically irrelevant. CERN's decision to go forward proved to be a wise one, especially after the SSC project went belly-up a few years later. Similarly, says Gerard 't Hooft, a Nobel Prize-winning physicist based at Utrecht University, even if China decides to build its new collider, "it would be wise for CERN to go forward with its plans too. We could end up with two machines, which would not be bad at all." The LHC and Tevatron overlapped for a couple of years, and science benefitted as a result. Many particle physicists would be happy to see a situation like that again.[42]

To some extent, the Chinese investigators pushing for a new collider are now waiting to see whether their government will finance the research proposal. But that doesn't mean the project is entirely on hold until a decision is laid down. During the interim, Wang says, his team is continuing to work on both the design of the machine and the relevant physics—the hope being that within a couple of years they'll turn their preliminary conceptual design report into a conceptual design report. Before they can reach that stage, many details will have to be ironed out.

The biggest, most obvious point to be settled concerns the size of the facility: Will the tunnel be 50 or 100 kilometers around, or somewhere in between? And will the proton collider, in turn, aim to reach 70 or 100 TeV? Right now, there are a lot of options, Wang says. "We need to narrow them down before we can start the real project."[43]

"China needs to be bold," recommends Henry Tye, who directs the Institute for Advanced Study at the Hong Kong University of Science and Technology. If the project goes forward, he urges China to build a big tunnel, at least 100 kilometers in circumference, because civil engineering costs are low in the country—much cheaper than those in many Western nations. And having a big tunnel means that the collision energy can keep going up as superconducting magnet technology improves. "What's the point of building the second best machine?" Tye asks. "In our field, second best is not good enough."[44]

Peskin agrees with that assessment, encouraging China to capitalize on its advantages by digging as large a tunnel as it can.[45] He believes that will be the most cost-efficient way of achieving 100 TeV collisions because with a smaller tunnel much more expensive magnets would be required to reach the same energy.

The international panel, which evaluated the preliminary conceptual design report, arrived at a somewhat similar conclusion, encouraging the CEPC-SPPC team to aim high—and to think big. If this project gets the go-ahead, the review committee wrote, it could be "the dominant machine for high-energy physics in the world for the next 50 years. Thus this machine cannot be a second-rate project."[46]

Arkani-Hamed has gotten almost the same message in his conversations with physicists from all over the world. They've told him that "50 kilometers would be fantastic but 100 kilometers would be a real game-changer, transforming the landscape of particle physics in the blink of an eye. This is a time to be daring and take that leap."[47]

Before leaping, however, there's still much to be done—identifying the chief technical concerns, working out a specific site plan, and addressing issues related to manpower. Given that the maximum beam energy of a hadron collider is directly proportional to the strength of the magnets that bend and focus those beams, the biggest technical challenges for the SPPC revolve around superconducting magnets. The most powerful magnets presently used at the LHC, which rely on niobium-titanium as the superconducting material, operate at magnetic field strengths of roughly 8 tesla. In order to reach 100 TeV, the SPPC would need durable and reliable 20-tesla superconducting magnets, most likely made out of a niobium-tin alloy—a material that is difficult to work with because of its brittleness but can generate significantly stronger magnetic fields. In tests with this material, Chinese physicists have reached 12 tesla in the laboratory, and researchers at the Lawrence Berkeley National Laboratory have exceeded 16 tesla.[48] "We want to reach 20 tesla, but we have 20 years to get there," says Wang.[49] On the other hand, technology doesn't just change by itself, he notes. "You need something to drive it. And in the past, big high-energy physics projects have done a lot to push technology."[50]

The people behind the LHC faced a similar situation in the early stages of planning. In 1983, two physicists then at CERN—Steve Myers and Wolfgang Schnell—concluded that 10-tesla magnets would be needed to achieve maximum beam energy. Because these magnets did not exist at that time, Myers and Schnell wrote, "such a project is not for the near future, and ... it should not be attempted before the technology is ready."[51]

In short, comments University College London physicist Jon Butterworth, a member of the ATLAS team at the LHC, "they didn't know how to build it, but they had to start doing research and development, or the technology might never exist. To build such machines, you don't just have to exploit cutting-edge technology and engineering, you have to develop new technologies." [52] In the same way, the China collider project is likely to directly or indirectly advance superconducting magnet technology, as well as a host of other technologies that would be incorporated into this ambitious undertaking.

In addition to questions of *how* to build the accelerator, one also must grapple with the matter of *where* to put it. At the same time that some researchers are looking into technology development, others are trying to pick out the best spot for a new Chinese collider. More than 14 possible sites were visited and surveyed, with a clear favorite emerging in the general vicinity of Qinghuangdao, a port city and beach resort 300 kilometers east of Beijing.

The site's biggest selling point is its favorable geology, as determined by an extensive survey of the area carried out by the Yellow River Engineering Consulting Company. The region is underlain by granite of the requisite thickness, which is thought to provide the lowest-cost option for tunnel building. Ground motion measurements have also shown the rock to be tectonically stable. In fact, other sites that were considered had ground motions that were 10 to 100 times higher, according to IHEP physicist Qing Qin. [53] But the tunnel cannot be moving perceptibly if physicists are going to achieve the necessary precision in their measurements. To confirm the geological suitability and identify any underground reservoirs, many more holes have to be drilled. "We want to make sure there are no surprises," Wang says. [54]

Another attribute is the Qinghuangdao site is its proximity to Beijing—a three-hour drive and a two-hour train ride. Next year, says Qin, a bullet train should start running that would make the trip in an hour. [55] Furthermore, the area offers an attractive destination that might appeal to foreign scientists. The air quality, compared to Beijing and other industrialized parts of China, is quite good, and the climate is relatively mild and temperate. With its many beaches, ranked among the country's finest, Qinghuangdao is sometimes called "China's summer capital." [56] The countryside—hilly, rural land reminiscent of parts of California and Italy—is home to some of China's leading vineyards and farms where various crops are cultivated. There appears to be plenty of room to build laboratories and housing for scientists, and local officials have been highly supportive of the project so far.

Despite the positive response from the regional government, says Wang, much work lies ahead in terms of securing the land and lining up the permits that will be needed before construction could possibly begin. A major effort is also required to win the support of the community at large. To this end, he and his colleagues are organizing workshops and symposia aimed at introducing the project to the general public. [57]

This will be part of a broader campaign to get people throughout China interested in high-energy physics and, perhaps even more critically, to encourage students to enter that field where there services will be desperately needed. Training enough qualified people may pose an even bigger challenge than developing the essential superconducting magnet technology, Arkani-Hamed says. China's physics program will have to grow at five percent per year for many years to support an operation of this scale. [58]

Project director Xinchou Lou concurs. "One of our most important jobs is to get young people excited enough to come into this field and ultimately become the scientific and engineering backbone of this project. Because, in the end, it's their project, and they're the ones who are going to have to carry it through." [59]

One of the best ways, historically, of drawing people into a particular area of science and technology is through big, high-profile projects that capture the public's imagination, like the Apollo "man-on-the-moon" missions. Of course, the Apollo program did more than put a man (or several men) on the moon. It gave a huge boost to the entire American scientific enterprise, and many feel that the new collider could play a similar role. "For China, this could be the man-on-the-moon project and more so," says Arkani-Hamed. "While the moon landings captivated the public, they didn't do much for science. The collider promises much more. It will be a crucial addition to our understanding of how the universe works." [60]

However, before an all-out recruitment program is initiated inside China, and efforts are simultaneously made to line up international partners, the CEPC-SPPC team must wait until they receive at least a preliminary green light from the government. But it is clear to Arkani-Hamed and practically all parties concerned "that if we are going to have a 100 TeV collider, we are going to need the entire world—and the talent of the entire world—involved." [61]

His colleague, IAS director Robbert Dijkgraaf, couldn't agree more. "The only way to have a successful project of this size is to work harmoniously with the whole scientific community. I can see China taking the initiative and putting in its own resources, while making sure that the rest of the world joins in." [62]

The prospect of foreign partners is one that Chinese scientists not only welcome but are becoming increasingly comfortable with. Over time, the country is gaining experience with international physics collaborations through the BESIII accelerator experiment and the Daya Bay and Juno neutrino experiments. About one-third of the physicists working on BESIII are non-Chinese, according to Wang. About half the physicists at Daya Bay are non-Chinese and more than half at Juno will be from institutions outside of China, he says. "These projects keep

The Center for Future High Energy Physics was inaugurated in Beijing on December 17, 2013 to map out the future of the field and to establish the detailed physics goals for a new collider. The director of this center, Nima Arkani-Hamed of the Institute for Advanced Study in Princeton (standing at left), unveiled the foundation for this center with IHEP director, Yifang Wang (standing at right). Behind them (from left to right) are Hesheng Chen, the former director of IHEP, and the Nobel Prize-winning physicist David Gross of the University of California, Santa Barbara. (Photograph courtesy of IHEP)

Xinchou Lou, a physicist who is coordinating the effort to build a large new collider in China, speaks at a 2013 IHEP conference on circular particle colliders. Lou is based at IHEP and at the University of Texas, Dallas. (Photograph courtesy of Xinchou Lou)

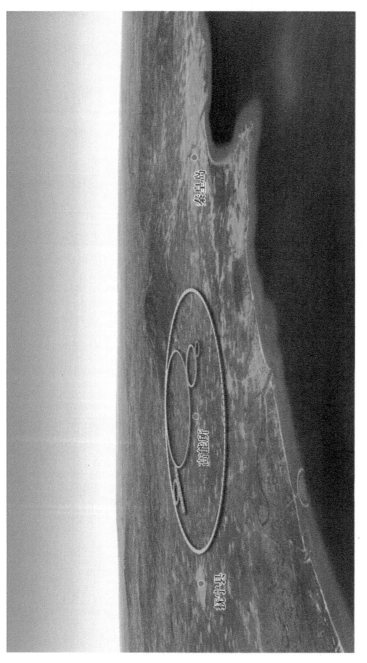

An aerial view of a potential site for the Super Proton–Proton Collider (SPPC), also referred to as the "Great Collider." The red dot at the far left designates the city of Funing, and the red dot at the far right the city of Qinhuangdao. The largest circle indicates the circumference of a 50-kilometer tunnel, and the smaller circles indicate boosters for the SPPC. (Image courtesy of IHEP)

getting more and more international, which is something that our scientists like and that our government is happy to see." [63]

When the time comes, there should be no problem getting the U.S. physics community to participate in a China-based accelerator program, says David Gross. "This would be a fantastic opportunity for U.S. physicists to take advantage of because our country does not have a comparable machine, nor does it have any plans to build one. We have a lot to contribute to China for the Great Collider"—including expertise in proton colliders and superconducting magnets, for instance—"but the project could also be of tremendous benefit to U.S. high-energy physics." [64]

Luciano Maiani views the involvement of a big country like China as a positive development in high-energy physics. The current arrangement, he notes, "of having everything based in Europe is not the healthiest situation for our field." [65] Establishing a balance between Europe, the United States, and Asia would put the field on better footing, adds Dijkgraaf. [66]

At this stage, it is too early to know how things will play out in China and the rest of the world in terms of high-energy physics. In the end, it may come down to a question of will, says Gross, who believes that "China has the will and the capacity to do this." The country's economic growth has put it in a position to take on projects like the planned collider that many other countries simply could not consider. [67]

Arkani-Hamed, who has been shuttling back and forth between the United States and China almost every month for the past year, is encouraged by what he's seen so far. "Given the boldness I see elsewhere in China, I think this project has a good chance. The most exciting sense I get is that this is a country that wants to do big things and thinks that big things are possible." [68]

Yifang Wang, his associate in this endeavor, is also hopeful that the key pieces, or "elements," are falling into place. First, the economy is still growing, which should leave China with the world's largest gross domestic product (GDP), possibly within about a decade. According to Wang, the cost of the CEPC and SPPC is a small fraction of the GDP—about 1/10,000—is no more than, and quite possibly less than the cost of the LHC and LEP and even of the diminutive Beijing Electron Positron Collider (BEPC), which was constructed in the 1980s. To him, those numbers suggest that the Great Collider will be in line with previous accelerators on a per-GDP basis and therefore not a great extravagance. [69]

Another prime motivating factor is that, within about a decade, operations at the BEPCII will be coming to a halt, leaving the nation in need of a high-energy physics project to fill the void. "So I think this is the moment," Wang says. [70]

Although he can't control the decisions made within the government of China, let alone in the rest of the world, Wang still believes "there is a significant possibility that this vision can be realized—enough of a chance that my colleagues and I are willing to pour everything into it, and do all we can, to try to make it happen." He has no difficulty getting motivated, despite the collider's uncertain future. "If you have a great opportunity to pull off something like this, you jump in and push for it with everything you've got," Wang says. "I don't think we need any other reasons beyond that." [71]

Chapter 6

The Most Amazing Spinoff of All

IN 1854, THE STORY GOES, WILLIAM GLADSTONE, the British Chancellor of the Exchequer (and future prime minister), asked Michael Faraday about the value of his experiments on electrical generation. Gladstone wanted to know whether this thing he was investigating, electricity, could be of any possible use. "I know not," Faraday allegedly replied, "but I wager that one day your government will tax it."[1]

There is some question as to the veracity of this attributed quote, its exact wording, and the year in which it was uttered. Yet the sentiment it expresses has surely been borne out by history. Faraday's purported statement demonstrated a confidence that basic research into physical phenomena like electricity—and basic research in general—would in all likelihood yield practical benefits down the road, though he could not spell out in advance exactly what those benefits would be. Prognostications are complicated by the fact that the payoff from such efforts may not be realized until several decades later.

The use of electricity for public street lighting was introduced in England and elsewhere in the world starting in the late 1870s, and the taxation of electricity sales soon followed—just as Faraday had "wagered." Moreover, the work that he and James Clerk Maxwell carried out in the mid-1800s culminated in a unified theory of electricity and magnetism, which helped pave the way for the "Electric Age," transforming life in the 20th century.

In 1905, while developing the theory of special relativity, Albert Einstein came up with his famous equation regarding the equivalence of mass and energy: $E=mc^2$. This formula provides the theoretical underpinning for nuclear fission, which was first demonstrated in 1938 by the Austrian-born physicists Lise Meitner and Otto Frisch[2]—and illustrated even more dramatically seven years later with the detonation of the first atomic bomb in a remote stretch of New Mexico desert. The formula also explains the energy behind nuclear fusion—the process that makes our sun shine and also gave us thermonuclear ("hydrogen") bombs and may one day provide us with a new means of generating electricity on a large scale, just as nuclear fission reactors do today.

The accuracy of the Global Positioning System (GPS), which became fully operational in 1995, owes to the theory of general relativity that Einstein first unveiled in 1915—70 years before this application came to light. Similarly, observes the physicist David Gross, "quantum mechanics, a highly abstract and theoretical development 100 years ago, is now the backbone of modern technology."[3] Without quantum mechanics we would not have transistors, integrated circuits, digital computers, cell phones, LEDs, lasers, Blu-ray players, and a vast array of other devices that we now take for granted—or quantum computers upon which many future hopes are now placed.

Spinoffs like this almost invariably come to the fore, says the physicist Steven Weinberg, "when we push back the frontiers of knowledge... The effort to make these discoveries forces us into a sort of technological and intellectual virtuosity that leads on to other applications."[4] In many cases these applications arise spontaneously, without any advance planning or even without deliberate measures taken to stimulate them. They just happen.

"Basic science generates the ideas of the future, which become applied science," notes Fermilab director Nigel Lockyer. "Without basic science you have no input to applied science, [and] the stream dries up." In other words, Lockyer adds, "Basic science is the future of everything."[5]

High-energy accelerators, designed over the decades to hasten progress in particle physics, routinely press the limits of human engineering capability. As such, they have consistently contributed to advances in a wide range of technical areas. To gain a sense of how pervasive this technology has been, one might note that even though accelerators were originally created to address questions in fundamental physics, only about 0.5% of the world's 25,000 accelerators are currently used for physics research, according to Oxford physicist Philip Burrows. The remaining 99.5% are used for biomedical, industrial, and other purposes.[6]

One important avenue for technological spinoffs has been medical imaging, where the particle detectors that were devised to track and identify particles with great precision have also proven to be a valuable means of scanning the human body. In the 1970s, for example, CERN physicists, working alongside researchers from a Geneva hospital, built one of the first positron emission tomography (PET) scanners—devices that are routinely used to look for disease and see how a person's internal tissues and organs are functioning.

It's safe to say that no one predicted this particular development in 1928, the year in which Paul Dirac wrote down an equation—drawing from both special relativity and quantum mechanics—which described the behavior of an electron moving at close to the speed of light. Dirac's equation had two solutions, calling for the existence of a particle identical to the electron in every way, except for having an opposite electric charge—the same particle that Carl Anderson discovered four years later and called a "positron."

In the 1980s, CERN physicists developed "imaging crystals" that glow when struck by certain particles. Although these crystals were initially intended for use in particle detectors, to map out the byproducts of a collision, they have since become common components of PET scanners. A decade later, CERN scientists showed how these same imaging devices could be utilized in the presence of strong magnetic fields—a finding that led to the advent of new medical tools that combined some of the features of both PET and magnetic resonance imaging (MRI). "What this story tells us is that ... basic science drives innovation," says CERN director-general Rolf Heuer, echoing words similar to those expressed by Fermilab's Lockyer. [7]

Commercial MRI machines were an outgrowth of the large-scale production of superconducting magnets for the Tevatron collider, according to University of Chicago physicist Young-Kee Kim, a former deputy director of Fermilab. [8] The Tevatron pioneered the use of niobium-titanium superconducting magnets, giving rise to a new industry in the process. This general class of magnets was later used at the LHC, which soon resulted in further progress on MRI machines. "This entire advance in medical imaging came out of something that only particle physicists would have pushed," says Duke physicist Ashutosh Kotwal. [9]

However, the most important spinoff to emerge from the LHC so far has been the invention of the World Wide Web. The Web grew out of a 1989 proposal by Tim Berners-Lee, who was then a software engineer at CERN. In his pitch to the laboratory, Berners-Lee suggested addressing "the problem of information access at CERN [through] the idea of linked information systems." [10] His boss at the time, the physicist Mike Sendall, found the proposal "vague but exciting," granting Berners-Lee permission to explore the idea further. [11] Although the system Berners-Lee came up with was originally meant to provide a way for LHC physicists located in different countries to share data, it has broadened tremendously from that single application. The general approach that he introduced has since grown into a network that spans the world, completely reshaping global communications in the process.

More than three billion people currently have access to the World Wide Web[12], where more than one trillion web pages are posted.[13] The Web is so ubiquitous today, and relied upon by so many people, observes Caltech physicist Sean Carroll, that "it's hard to imagine the world without it. Nobody ever suggested funding CERN because some day they would invent the Web; it's just a matter of putting smart people into an intense environment with daunting technological challenges, and reaping the benefits from what comes out."[14]

Grid computing is another innovation that started at the LHC, building on the technology of the World Wide Web, and spread from there. This approach has been necessary for handling the enormous volume of data generated by the experiment, which owes to the fact that the LHC can produce up to one billion particle collisions each second.[15] Researchers are therefore obliged to sift through up to 30 petabytes—or some 30 billion megabytes—of data each year to determine whether any especially noteworthy physics has taken place.[16] Having neither the computational capacity nor the money to process all this data on site, CERN established the Worldwide LHC Computing Grid (WLCG) in 2002 to share the data-crunching responsibilities. The WLCG is the world's largest computing grid, which connects computers and storage systems in more than 170 centers spread across 41 countries, thus giving more than 8,000 physicists almost real-time access to LHC data.[17]

But the implications of the "Grid" extend well beyond the LHC. Such an approach to computing, says physicist Joseph Incandela, former spokesperson for the CMS experiment, "will revolutionize many areas of science," which have also been struggling with the challenges of handling immense quantities of data.[18]

Some of the benefits stemming from a facility like the LHC are more regional in nature, conferring advantages to Geneva and Switzerland, as well as to France, where half the machine is located. New jobs are created, with people engaged in a benign, non-polluting activity. Scientists flock to the area in large numbers and spend money there too, says former CERN director-general Luciano Maiani, "so it's been a big boost to the region and the local economy."[19] The LHC, Maiani adds, has been a great success story, but not only for the dazzling scientific findings that have been yielded to date. "It is a story of money invested with great returns to the governments in terms of image, technology, and the training of young people."[20]

China could stand to get a similar boost should the Great Collider come to fruition. "Lots of things can benefit from hosting a big accelerator project," says U.C. Berkeley physicist Hitoshi Murayama. "The country would open up, bringing in new talent, new resources, and creating more industries." This same phenomenon, he adds, has already been witnessed in and around CERN, where the establishment of the world's largest physics research center has been a tremendous boon to the region.[21]

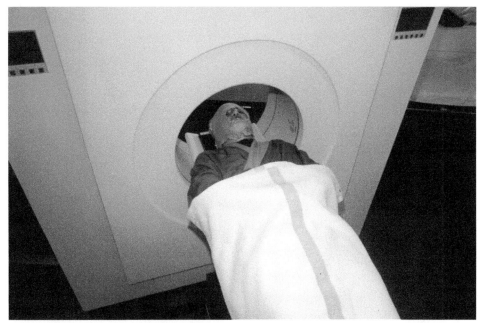

This Positron Emission Tomography (PET) scanner was developed, to a large extent, at CERN in the mid-1970s. (Photograph courtesy of CERN)

This crystal, of a type known as a scintillator, glows when high-energy charged particles or photons pass through it. Although scintillator crystals were originally developed at CERN for use in particle detectors, they have since been incorporated into PET scanners and other medical imaging devices. (Photograph courtesy of CERN)

A superconducting (dipole) magnet developed for the Tevatron particle collider. (Photograph courtesy of Fermilab)

Tim Berners-Lee invented the World Wide Web while working at CERN to create a computer system capable of handling and distributing the vast amount of data expected from the Large Hadron Collider. The world's first website, which Berners-Lee built at CERN, went online in 1991. (Photograph courtesy of CERN)

Many believe that building the next-generation collider would have the potential not only to promote high-energy physics research in China, but to lift up *all* science in China. However, in weighing the prospective returns on investment, says Henry Tye, high-energy physics stands out from virtually every other scientific discipline because the topflight facilities tend to be one-of-a-kind. "Even if China were to spend more than the United States in some field of science and engineering other than high-energy physics, U.S. professors would still do their research in the U.S. But if a giant collider were built here, almost every high-energy physics professor [from the U.S. and elsewhere] would have some incentive to move his or her research to China. Thus the worldwide impact of a collider is much bigger than if the money were put into some other area of science." [22]

A good example of this effect can be seen at the LHC, where as many as 10,000 scientists and engineers have been assembled to carry out research at any one time. When you factor in the large number of scientists involved and the number of years of research afforded by a new facility—typically a couple of decades—the high-energy research budget is actually no higher than research funding in other experimental fields, Tye contends. But the payoff can be much bigger. [23]

If China were to pursue a large-scale project in this area, one result would likely be an enhancement of national prestige, with the country becoming a leader in the field of high-energy physics and perhaps eventually becoming the world center for such research. Improved international relations may be the most important consequence of all. "This would be a scientific collaboration between the East and West on a scale that has never happened before," says Nima Arkani-Hamed. "Anything that can foster cooperation and friendship is good. When you get 10,000 people working together, interacting together, it's an important way of building bridges. There are so many aspects to this project that have the potential for so much good." [24]

If China opens it doors and becomes host to a huge international collaboration in basic research, the country will be richer for it, owing to all the interactions that ensue—among particles and among people. In some periods of its history, by contrast, China has been closed and isolated, and its science has lagged behind that in other parts of the world, says Yifang Wang. "Having an open-door policy is vital. It can benefit the whole society, energizing scientific research while bringing more hope for the future." [25]

As discussed before, a big collider project is almost certain to bring with it a big technological upside. High-energy physics in China, which has so far proceeded on a modest scale, has already fostered the creation of the first Chinese website (at IHEP), the forging of the first internet connection with the outside world, and the introduction of grid computing. [26] The BESIII detector at the

BEPCII pioneered the use of superconducting magnets unlike any previously fabricated in China. IHEP has also developed the nation's first micro-PET and micro-CT (computerized tomography) scanners.

The Great Collider, of course, would assume a scope and magnitude far beyond anything undertaken in Chinese physics before, strengthening the country's technological capacities in high-temperature superconductors, radiation-resistant materials, advanced instrumentation, high-speed computation, data processing and storage, communications networks, and so forth. "All of these would go along with a large-scale accelerator," says Weiren Chou, a Fermilab physicist and an advisor to the CEPC-SPPC program. "All of these will be frontier technologies. You'll be pushing the envelope to the limit on multiple fronts." [27]

Superconducting magnets, in particular, will get a big push from the project. "China is pinning their hopes on new niobium-tin superconductors, which could spawn a brand-new industry and possibly support future applications like maglev trains," says Kotwal. [28]

It is impossible to know what may come in the way of spinoffs, although some will surely materialize. The Higgs boson serves as a cornerstone for one of the most ambitious and successful theories in the history of physics, the Standard Model, yet no one has a clue as to whether someday this particle might hold any "practical" value. "Believe it or not, no one knew what the electron was good for when it was discovered," says Lisa Randall. The fantastic commercial potential of quantum mechanics, which has since proved to be critical for the semiconductor and electronics industry, was not evident at the dawn of the 20th century, when this idea was just starting to take hold. The benefits that come from cutting-edge discoveries are not always straightforward, nor are they obvious. What we have seen instead, Randall maintains, is that "societies accompany advanced science with advanced education and generally with a thriving economy that derives directly and indirectly from scientific developments." [29]

The societal advantages derived in this way can be extraordinarily important, yet they are difficult to quantify up-front and are almost never the primary drivers of an endeavor like the LHC, nor in all likelihood of the machine that may someday replace it. "Our greatest motivation is our love of fundamental science," says Joe Incandela, and the chance "to contribute to a tradition that is, to some extent, immortal." [30]

The immortality of high-energy physics, however, should not be taken for granted. Without a new accelerator on which investigators can undertake new experiments and practice their discipline, the whole field could conceivably dry up and die—a fate that almost every researcher desperately hopes to avoid. "The scientific revolution has been going on for many centuries, and we want to make sure that effort continues," maintains IAS director Robbert Dijkgraaf. "We want to make sure that the 21st century isn't the century we stop pushing boundaries." [31]

This brings us squarely to the heart of the matter. Particle physicists are not, as their job title might suggest, solely preoccupied with finding more and more particles. In a way, they're after something bigger—something that's been a part of humanity practically forever—a deep-seated desire to understand the world around them at the most basic level, without any thought given to material reward or eventual applications. "Why do humans do science?" asks Stanford physicist Savas Dimopolous, a key contributor to supersymmetry theory, among other areas. For no particular reason, he suggests. "The things that are least important for our survival are the very things that make us human." [32]

Sam Ting offers a somewhat simpler answer to the question regarding the impetus behind this sometimes esoteric research: "It's curiosity that drives a physicist forward to search for the unknown," he explains. [33]

The search to which Ting refers lies at the center of the scientific enterprise. Physicists, of course, always hope to obtain new results. They're anxious to discover new particles, identify new forces or symmetries, and find the answers to their questions. But they also understand that most answers they come by will be of a provisional character—resolving some puzzles while underscoring our ignorance in other areas—and always subject to revision as new findings come to light.

In science, Dijkgraaf explains, "you are at best a link in a long chain." [34] It is unrealistic to expect that the next particle you discover will clear up all remaining mysteries—or that the next theory you draw up will be the "final theory," the one that explains everything. More likely than not, it will help us ask new questions that we hadn't thought of before. And even if this new theory is any good, we cannot expect it to stand forever, in its original form, without addition, modification, or extension. Instead, it will merely carry us a bit further down the road, serving as a temporary way station on this long journey until we've mustered the resources necessary to take the next step ahead.

The essential point is to keep this great tradition alive—both to sustain it and constantly reinvigorate it—by participating full-on in the hunt for knowledge. Sometimes, to be sure, the going may get tough. Sometimes the obstacles—theoretical, technical, financial, logistical, political, and otherwise—may be truly daunting, making it hard to find a path forward. But we'll never be content to stop and say, "That's it, we've learned enough." Because the fact is, we'll always want to know more.

That yearning, as Dimopoulos has said, may be the key trait that defines us as humans. It is also the principal factor that will compel us, as it has time and again, to probe deeper into the mysteries of nature. It got us to where we are today, with the LHC standing as a towering monument to curiosity, [35] even if that "towering" structure is almost entirely underground. And with any luck, along with a strong sense of initiative and resolve, that same spirit will carry us to the next stage of this grand exploration.

EPILOGUE

What Lies Beneath

SHANHAI PASS, since antiquity referred to as the "First Pass Under Heaven," is strategically located between the Yanshan Mountains to the north and the Bohai Sea to the south. The pass, whose name, logically enough, is derived from the words for mountain ("Shan") and sea ("hai'), lies only about 15 kilometers from Qinhuangdao—a popular beach vacation spot and bustling port that is home to more than two million people, still a fairly modest-sized town by Chinese standards.

Throughout history, Shanhai had been one of the most heavily fortified passes in China—a focal point for national forces attempting to keep outside aggressors from making their way through the mountains and thereby gaining access to the capital cities of Beijing in the east and, in ancient times, Chang'an (Xi'an) to the west.

Today, the greatest threat to the Great Wall comes not from hostile intruders seeking to get across it but rather from the millions of tourists who visit this far-reaching monument each year. Sometimes these overly zealous guests take bricks and stones as souvenirs, in violation of local ordinances, but even the law-abiding types can inadvertently cause damage by contributing to the sheer volume of foot traffic and to the erosion that stems from that.

When the weather is nice, thousands of people a day visit the Shanhai portion of the wall, which is one of the better preserved stretches of a structure that spans, in its entirety, a significant portion of northern China. The walled city of Shanhaiguan, built in the 1300s during the Ming Dynasty, ranks among the favorite destinations in this area. Other places of interest nearby include the Temple of Mengjiangu (located at the foot of the Great Wall), Yansaihu Lake (at the base of Yanshan Mountain), Changshou (also known as "Longevity") Mountain, and Laolongtu, or "Old Dragon's Head," the Great Wall's easternmost terminus, as well as the spot where the Wall literally plunges into the sea.

The odds are that almost none of the sightseers exploring these historic sites are cognizant of the fact that, a decade or so from now, a Great Collider may be accelerating particles—electrons, positrons, and eventually protons and ions—at "relativistic" speeds, right beneath their feet—though a rocky floor, up to 100 meters thick, will provide more than adequate shielding from the high-voltage goings-on below.

The new accelerator, unlike the Great Wall, will never come into direct contact with the Pacific Ocean, keeping a distance of at least a few kilometers from the world's largest body of water. But it will, nevertheless, tap into something equally vast and deep, if not more so—the boundless reservoir of natural law.

One of the world's major tourism attractions, the Great Wall is no longer meant to play any defensive role. And the Great Collider, as stated before, was conceived from the outset with the opposite purpose in mind. Rather than trying to keep people out of China, the collider hopes to lure them in—particularly physicists and engineers—to take part in a rousing research collaboration, which seeks to expand the borders of our understanding at many points, and in many directions, along the leading edge of scientific inquiry.

The proposed accelerator will be an expensive project, to be sure, but it cannot be justified on military or national security grounds. Nor will it make any contributions towards the defense of China, apart from the safety that comes from enhanced international cooperation, camaraderie, and good will. But it will—as the physicist Robert Wilson once said about the Tevatron collider—help make it a country that is worth defending, and the same holds true for human society at large.

If it is erected someday, this mammoth high-energy physics facility will not be built for the glory of any one nation but rather for the advancement of science around the globe. And if the project goes forward, as many intellectually driven people hope it does, this ambitious venture will uphold, support, and defend to the fullest the most human of pursuits—that of seeking knowledge for its own sake, simply because we're curious. It will come to pass because we have the wit and means to make progress on some of the most profound and timeless questions that mere mortals have dared to ask.

The Great Wall of China meets the Pacific Ocean at Laolongtou, or "Old Dragon's Head," near the city of Qinhuangdao. (Photograph courtesy of China Highlights)

NOTES

Notes to Prologue: The Next Great Wall

1. Debra Bruno. "Getting the Great Wall to Talk." *Wall Street Journal.* July 12, 2012.

2. Jim Yardley. "Saving the Great Wall From Being Loved to Death." *New York Times,* November 26, 2006.

3. UNESCO. "The Great Wall." World Heritage List, http://whc.unesco.org/en/list/438

4. "The Great Wall of China." *National Geographic* (online). http://video.nationalgeographic.com/video/exploreorg/china-great-wall-eorg

5. "After the Higgs Discovery: Where is Fundamental Physics Going?" (Symposium). Tsinghua University, Beijing, February 23, 2014.

6. Frank Wilczek. "'National greatness' versus real national greatness." *Science News.* October 11, 2008, p. 32.

7. Henry Tye, Hong Kong University of Science and Technology (e-mail correspondence with the author), April 4, 2015.

Notes to Introduction: A Clarion Call

1. "On the Shoulders of Giants" (online). http://isaacnewton.org.uk/essays/Giants

2. "Particle Fever" (documentary). Directed by Mark Levinson, 2013.

3. Interview. Nima Arkani-Hamed. IHEP. August 8, 2014.

4. Nima Arkani-Hamed. Large Hadron Collider Physics Conference. Columbia University. June 6, 2014.

5. Ibid.

6. David Gross. "After the Higgs Discovery: Where is Fundamental Physics Going?" Symposium, Tsinghua University, February 23, 2014.

7. Edward Witten. "After the Higgs Discovery: Where is Fundamental Physics Going?" Symposium, Tsinghua University, February 23, 2014.

8. Luciano Maiani. "After the Higgs Discovery: Where is Fundamental Physics Going?" Symposium, Tsinghua University, February 23, 2014.

9. Hitoshi Murayama. "After the Higgs Discovery: Where is Fundamental Physics Going?" Symposium, Tsinghua University, February 23, 2014.

10. Nima Arkani-Hamed. "After the Higgs Discovery: Where is Fundamental Physics Going?" Symposium, Tsinghua University, February 23, 2014.

Notes to Chapter 1

1. Erik Gregersen. *The Britannica Guide to the Atom.* (New York: Britannica Educational Publishing, 2011), p. 32.

2. Dan Hooper. *Nature's Blueprint.* (New York: Harper Collins, 2008), pp. 179-180.

3. Henry C. Plotkin. *Evolutionary Worlds Without End.* (New York: Oxford University Press, 2010), p. 1.

4. American Institute of Physics (AIP), Center for History of Physics, from Rutherford's Nuclear World. "Atop the Physics Wave: Rutherford Back in Cambridge, 1919-1937". http://www.aip.org/history/exhibits/rutherford/sections/atop-physics-wave.html

5. "Cockcroft's subatomic legacy: splitting the atom." *CERN Courier* (online). November 20, 2007. http://cerncourier.com/cws/article/cern/31864.

6. "Ernest Lawrence's Cyclotron: Invention for the Ages." *Berkeley Lab Science Articles Archive* (online). http://www2.lbl.gov/Science-Articles/archive/early-years.html

7. Edward Witten. "Matter matters." *The New Republic.* December 29, 1997, p. 16.

8. Steven Weinberg. *Dreams of a Final Theory.* (New York, Vintage Books, 1992), p. 266.

9. Edward Witten. "After the Higgs Discovery: Where is Fundamental Physics Going?" Symposium, Tsinghua University, February 23, 2014.

10. Frank Close. *The Infinity Puzzle.* (New York: Basic Books, 201), p. 314.

11. Arthur Fisher. "A Different Kind of Gravity." *Mosaic.* 17 (Winter 1986/7), p. 20.

12. Marcia Bartusiak. "Science and Technology: Who Ordered the Muon?" *New York Times* (online). September 27, 1987.

13. Lynn Yarris. "The Golden Anniversary of the Antiproton." *Berkeley Lab Science Articles Archive* (online).
    http://www2.lbl.gov/Science-Articles/Archive/sabl/2005/October/01-antiproton.html

14. T. D. Lee and C. N. Yang. "Question of Parity Conservation in Weak Interactions." *Physical Review.* 104 (October 1, 1956), pp. 254-258.

15. "C S Wu—First Lady of physics research." *CERN Courier* (online). November 27, 2012.
    http://cerncourier.com/cws/article/cern/51556.

16. William Dicke. "Chien-Shiung Wu, 84, Dies; Top Experimental Physicist." *New York Times* (online). February 18, 1997.

17. Michael Peskin, SLAC (e-mail communication with the author), May 25, 2015.

18. "Chien-Shiung Wu, 84, Dies."

19. Luis W. Alvarez. "Recent Developments in Particle Physics. *Science.* 165 (September 12, 1969), pp. 1071-1091.

20. Michael Creutz. "Yang-Mills Fields and the Lattice." *50 Years of Yang-Mills Theory.* Edited by Gerard 't Hooft. (Singapore: World Scientific, 2005), p. 359.

21. Gloria Lubkin. "Friedman, Kendall, and Taylor Win Nobel Prize for First Quark Evidence." *Physics Today.* January 1991, p. 18.

22. Ibid., p. 19.

23. Elizabeth Kolbert. "Crash Course." *The New Yorker*, Annals of Science (online). May 14, 2007.
    http://www.newyorker.com/magazine/2007/05/14/crash-course

24. Charles Pierce. "Samuel Ting's space odyssey." *Boston Globe Magazine.* April 10, 2011.

25. Suzanne Jacobs. "The Revolution That Shook Particle Physics." *MIT Technology Review* (online). October 21, 2014.
    http://www.technologyreview.com/article/531396/
    the-revolution-that-shook-particle-physics/

26. "Nobel Voices Video History Project, 2000-2001." Samuel Ting. Archives Center, National Museum of American History. June 29, 2000.

27. Carlos I. Calle. "The Universe—Order Without Design." (Amherst, New York: Prometheus Books, 2009), p. 114.

28. Charles Pierce. "Samuel Ting's space odyssey." *Boston Globe Magazine.* April 10, 2011.

29. Martin L. Perl. "The Discovery of the Tau Lepton." SLAC-PUB-5937, September 1992.

30. John Ellis. "Those were the days: discovering the gluon." *CERN Courier* (online). July 15, 2009.
http://cerncourier.com/cws/article/cern/39747

31. Michael Albrow. "Particle physics has a big future." *New Scientist.* April 12, 1979, p. 100.

32. Ilka Flegel and Paul Söding. "Twenty-five years of GLUONS." *CERN Courier* (online). November 12, 2004.
http://cerncourier.com/cws/article/cern/29201

    John Ellis. "Those were the days: discovering the gluon." *CERN Courier* (online). July 15, 2009.
    http://cerncourier.com/cws/article/cern/39747.

33. Frank Wilczek. "QCD Made Simple." *Physics Today.* August 2000, p. 22.

34. Frank Close. *The Infinity Puzzle.* (New York: Basic Books, 2011), pp. 80-91.

35. Steven Weinberg. "The Making of the Standard Model." *Arxiv.org.* January 3, 2004.
http://arxiv.org/abs/hep-ph/0401010

36. Ibid.

37. Norbert Straumann. "Wolfgang Pauli and Modern Physics." October 14, 2008.
http://arxiv.org/pdf/0810.2213v1.pdf

38. The Nobel path to a unified electroweak theory." *CERN Courier* (online). December 7, 2009.
http://cerncourier.com/cws/article/cern/41013

39. The Power of the Weak Force." *CERN Press Office* (online). September 2, 2003.
https://cern-discoveries.web.cern.ch/cern-discoveries/Story/WelcomeStory.html

40. Peter Renton. "The W boson weighs in." *Physics World.* January 2003, pp. 29-34.

41. Gordon Kane. "The Dawn of Physics Beyond the Standard Model." *Scientific American.* June 2003, p. 70.

42. Gerard 't Hooft. "After the Higgs Discovery: Where is Fundamental Physics Going?" Symposium, Tsinghua University, February 23, 2014.

43. Ann Finkbeiner. "Looking for Neutrinos, Nature's Ghost Particles." *Smithsonian Magazine* (online). November 2010.

44. "This Month in Physics History. July 21, 2000: Fermilab announces first direct evidence for tau neutrino." *APS News* (online) 20, July 2011. http://www.aps.org/publications/apsnews/201107/physicshistory.cfm

45. Dan Hooper. *Nature's Blueprint.* (New York: Harper Collins, 2008), pp. 125-126.

46. Matt Strassler. "The Higgs FAQ 2.0." *Of Particular Significance* (blog). October 12, 2012. http://profmattstrassler.com/articles-and-posts/ the-higgs-particle/the-higgs-faq-2-0/

47. Gloria Lubkin. "Friedman, Kendall, and Taylor Win Nobel Prize for First Quark Evidence." *Physics Today.* January 1991, p. 20.

Notes to Chapter 2

1. Gerard 't Hooft. "After the Higgs Discovery: Where is Fundamental Physics Going?" Symposium, Tsinghua University, February 23, 2014.

2. Matt Strassler (interview). Harvard University, December 16, 2014.

3. Kathryn Jepsen. "Famous Higgs analogy, illustrated." *Symmetry* (online). September 6, 2013. http://www.symmetrymagazine.org/article/september-2013/ famous-higgs-analogy-illustrated

4. David J. Miller. "A quasi-political Explanation of the Higgs Boson; for Mr. Waldegrave, UK Science Minister 1993." http://www.hep.ucl.ac.uk/~djm/higgsa.html.

5. Kathryn Jepsen. "Ten Things you many not know about the Higgs boson." *Symmetry* (online). March 1, 2012. http://www.symmetrymagazine.org/article/march-2012/ ten-things-you-may-not-know-about-the-higgs-boson

6. Brian Greene. *The Fabric of the Cosmos.* (New York, Alfred A. Knopf, 2004), pp. 252-264.

7. Lisa Randall. *Higgs Discovery.* (New York, Ecco, 2013), pp. 76-77.

8. Nigel Lockyer. "Massive thoughts." *Symmetry.* April 24, 2014.

9. Brian Greene. *The Fabric of the Cosmos.* (New York, Alfred A. Knopf, 2004), p. 262.

10. Philip Ball. "Nuclear masses calculated from scratch." *Nature.* November 20, 2008.

11. Juan Maldacena. "The symmetry and simplicity of the laws of physics and the Higgs boson." *Arxiv.org.* October 24, 2014.
http://lanl.arxiv.org/abs/1410.6753

12. Gerard 't Hooft. "After the Higgs Discovery: Where is Fundamental Physics Going?" Symposium, Tsinghua University, February 23, 2014.

13. Nima Arkani-Hamed. "After the Higgs Discovery: Where is Fundamental Physics Going?" Symposium, Tsinghua University, February 23, 2014.

14. "R.R. Wilson's Congressional Testimony, April 1969." Fermilab History and Archives Project.
http://history.fnal.gov/testimony.html

15. Dennis Overbye. "Physicists Inch Closer to Proof of Elusive Particle." *New York Times* (online). July 2, 2012.

16. Gary Taubes. "The Supercollider: How Big Science Lost Favor and Fell." *New York Times.* October 26, 1993.

17. Steven Weinberg. *Dreams of a Final Theory.* (New York, Vintage Books, 1992), p. 278.

18. Gary Taubes. "The Supercollider: How Big Science Lost Favor and Fell." *New York Times.* October 26, 1993.

19. Harry Lustig. "To advance and diffuse the knowledge of physics: An account of the one-hundred-year history of the American Physical Society." *American Journal of Physics.* 68 (July 2000), p. 632.

20. Lyndon Evans. "The Large Hadron Collider from Conception to Commissioning A Personal Recollection. *Reviews of Accelerator Science and Technology.* 3 (December 2, 2010), p. 261.

21. Dennis Overbye. "Asking a Judge to Save the World and Maybe a Whole Lot More." *New York Times* (online). March 29, 2008.

22. "LHC to Restart in 2009." *CERN Press Office* (online). December 5, 2008.
http://press.web.cern.ch/press-releases/2008/12/lhc-restart-2009

23. Lyndon Evans. "The Large Hadron Collider from Conception to Commissioning A Personal Recollection. *Reviews of Accelerator Science and Technology.* 3 (December 2, 2010), p. 279.

24. Sean Carroll. *The Particle at the End of the Universe.* (New York: Dutton, 2012), pp. 88-90.

25. Melissa Franklin. "Putting the Higgs Boson in Its Place" (lecture). Harvard University, October 30, 2014.

26. Lisa Randall. "Foreword." *Most Wanted Particle* by Jon Butterworth. (New York: The Experiment, 2015), p. xiii.

27. "Particle Fever" (documentary). Directed by Mark Levinson, 2013.

28. Sean Carroll. *The Particle at the End of the Universe.* (New York: Dutton, 2012), p. 99.

29. "What is ATLAS?" *ATLAS Experiment* (online). http://atlas.ch/what_is_atlas.html

30. "Why is CMS So Big?" Compact Muon Solenoid Experiment at CERN's LHC (website). http://cms.web.cern.ch/news/why-cms-so-big

31. "LHC Machine Outreach." http://lhc-machine-outreach.web.cern.ch/lhc-machine-outreach

32. Matt Strassler. "Will the Higgs Boson Destroy the Universe." *Of Particular Significance* (blog), September 10, 2014. http://profmattstrassler.com

33. Matt Strassler. "A lightweight Standard Model Higgs Particle." *Of Particular Significance* (blog), December 5, 2011. http://profmattstrassler.com

34. Matt Strassler (interview). Harvard University, December 16, 2014.

35. Matt Strassler. "Why is it Hard to Find the Higgs Particle?" *Of Particular Significance* (blog). Undated. http://profmattstrassler.com

36. Lisa Randall. *Higgs Discovery.* (New York, Ecco, 2013), p. 21.

37. Matthew Schwartz. "Discoveries at CERN—Panel Discussion." Harvard University, October 23, 2014.

38. Matt Strassler. "Higgs Discovery: The Data." *Of Particular Significance* (blog), July 6, 2012. http://profmattstrassler.com

39. Particle Fever" (documentary). Directed by Mark Levinson, 2013.

40. Sean Carroll. *The Particle at the End of the Universe.* (New York: Dutton, 2012), p. 184.

41. Particle Fever" (documentary). Directed by Mark Levinson, 2013.

42. Celeste Biever. "Celebrations as Higgs boson is finally discovered." *New Scientist* (online). July 4, 2012. http://www.newscientist.com

43. Particle Fever" (documentary). Directed by Mark Levinson, 2013.

44. John Ellis. "Theory Summary and Prospects." *Arxiv.org.* September 15, 2014. http://arxiv.org/pdf/1408.5866.pdf

45. "Particle Fever" (documentary). Directed by Mark Levinson, 2013.

46. "The Large Hadron Collider." *CoEPP* (online).
    http://www.coepp.org.au/large-hadron-collider

47. Sean Carroll. *The Particle at the End of the Universe*. (New York: Dutton, 2012), p. 79.

48. Joseph Incandela. "After the Higgs Discovery: Where is Fundamental Physics Going?"
    Symposium, Tsinghua University, February 23, 2014.

49. Edward Witten. "After the Higgs Discovery: Where is Fundamental Physics Going?"
    Symposium, Tsinghua University, February 23, 2014.

50. Nima Arkani-Hamed. "After the Higgs Discovery: Where is Fundamental Physics
    Going?" Symposium, Tsinghua University, February 23, 2014.

51. Sean Carroll. *The Particle at the End of the Universe*. (New York: Dutton, 2012), p. 186.

52. Cian O'Luanaigh. "New results indicate that new particle is a Higgs boson." *CERN*
    (online). March 14, 2013.
    http://home.web.cern.ch/about/updates/2013/03/
    new-results-indicate-new-particle-higgs-boson

53. Robbert Dijkgraaf, Institute for Advanced Study (telephone interview with the author),
    April 20, 2015.

Notes to Chapter 3

1. Nima Arkani-Hamed. "Beyond the Standard Model theory." *Physica Scripta*. T158
    (December 2013), pp. 014023-24.

2. John Ellis. "The Beautiful Physics of LHC Run 2." *Proceedings of Science*.
    December 8, 2014.
    http://arxiv.org/abs/1412.2666

3. Joseph Incandela. "After the Higgs Discovery: Where is Fundamental Physics Going?"
    Symposium, Tsinghua University, February 23, 2014.

4. Steven Weinberg. "What We Do and Don't Know." *New York Review of Books*.
    November 7, 2013.

5. Sergio Bertolucci. Second Annual Large Hadron Collider Physics Conference,
    Columbia University, June 6, 2014.

6. Matt Strassler. "Why the Higgs Particle Matters." *Of Particular Significance (blog)*.
    July 2, 2012. http://profmattstrassler.com

7. G. Aad, *et al.* "Combined Measurement of the Higgs Boson Mass in *pp* collisions." The
    Atlas and CMS Collaborations. *arXiv:1503.07589v1*. March 26, 2015.

8. CERN Press Office. "LHC experiments join forces to zoom in on the Higgs boson." *CERN Press Release.* March 17, 2015, http://press.web.cern.ch/press-releases/2015/03/ lhc-experiments-join-forces-zoom-higgs-boson

9. "China super-collider plan for 'God particle' studies." *ShanghaiDaily.com.* February 27, 2014, http://www.shanghaidaily.com/national/ China-supercollider-plan-for-God-particle-studies/shdaily.shtml

10. Nima Arkani-Hamed, Institute of Advanced Studies (interview in Cambridge, Mass.). December 2, 2014.

11. "LHC experiments are back in business at a record new energy." *CERN Press Office* (online). June 3, 2015. http://press.web.cern.ch/press-releases/2015/06/ lhc-experiments-are-back-business-new-record-energy

12. Elizabeth Gibney. "LHC 2.0." *Nature.* 519 (March 12, 2015), p. 142.

13. Ashley Wenners-Herron and Kathryn Jepson. "What's Next for the Large Hadron Collider?" *Symmetry* (online). February 4, 2013. http://www.symmetrymagazine.org/article/february-2013/ whats-next-for-the-large-hadron-collider

14. Luciano Maiani. "After the Higgs Discovery: Where is Fundamental Physics Going?" Symposium, Tsinghua University, February 23, 2014.

15. Nima Arkani-Hamed. The Second Annual Large Hadron Collider Physics Conference. Columbia University. June 6, 2014.

16. Rebecca Boyle. "CERN Physicists to Build Longest-Ever Linear Particle Accelerator." *Popular Science* (online). July 19, 2010.

17. R.A. "The Q&A: Brian Greene – Life after the Higgs." *The Economist* (online). July 19, 2012. http://www.economist.com/blogs/babbage/2012/07/qa-brian-greene

18. Yifang Wang, IHEP (interview with the author at Harvard University), December 2, 2014.

19. Lisa Randall. *Warped Passages.* (New York, Ecco, 2005), p. 253.

20. Nima Arkani-Hamed, Savas Dimopoulos, and Georgi Dvali. "The Universe's Unseen Dimensions. *Scientific American.* August 2000, p. 62.

21. Lisa Randall. *Higgs Discovery.* (New York, Ecco, 2013), p. 94.

22. Lisa Randall. *Warped Passages.* (New York, Ecco, 2005), pp. 243, 250.

23. Don Lincoln. *The Large Hadron Collider.* (Baltimore, Johns Hopkins University Press, 2014), p. 167.

24. Edward Witten. "Matter matters: why we need supercolliders." *The New Republic.* December 29, 1997, pp. 16-17.

25. Gordon Kane. *Supersymmetry,* (Cambridge, Mass, Perseus Publishing, 2000), pp. 5-6.

26. Dan Hooper. *Nature's Blueprint.* (New York, Harper Collins, 2008), pp. 109-110.

27. Gordon Kane. *Supersymmetry,* (Cambridge, Mass, Perseus Publishing, 2000), p. 68.

28. John Ellis, CERN (telephone interview with author), January 5, 2015.

29. Joseph Lykken and Maria Spiropulu. "Supersymmetry and the Crisis in Physics." *Scientific American.* May 2014, p. 34-39.

30. Gordon Kane. *Supersymmetry,* (Cambridge, Mass, Perseus Publishing, 2000), pp. 55-56.

31. John Ellis (telephone interview with author), January 5, 2015.

32. Dan Hooper. *Nature's Blueprint.* (New York, Harper Collins, 2008), p. 193.

33. M. E. Peskin. "Will there by Supersymmetry at the ILC?" (talk presented at CERN workshop, *Implications of LHC results for TeV-scale physics.* July 2012.

34. Michael Peskin, SLAC (e-mail communication with the author), May 25, 2015.

35. Matt Strassler, Harvard University (interview with the author at Harvard University), December 16, 2014.

36. Masahiro Morii, Harvard University (interview with the author at Harvard University), December 19, 2014.

37. Jill Sakai. "Heart of the Matter." *University of Wisconsin-Madison News* (online). http://www.news.wisc.edu/on-wisconsin/heart-of-the-matter

38. Masahiro Morii (interview with the author at Harvard University), December 19, 2014.

39. Alvin Powell. "Back into the dark." *Harvard Gazette.* December 16, 2014, http://news.harvard.edu/gazette/story/2014/12/back-into-the-dark

40. Nima Arkani-Hamed (interview with the author in Cambridge, Mass.). December 2, 2014.

41. Nima Arkani-Hamed (interview with the author at IHEP in Beijing). August 8, 2014.

42. Nima Arkani-Hamed. After the Higgs Discovery: Where is Fundamental Physics Going?" Symposium, Tsinghua University, February 23, 2014.

43. Edward Witten. "Foreword." appearing in *Supersymmetry* by Gordon Kane, (Cambridge, Mass, Perseus Publishing, 2000), p. xiii.

44. Frank Close. *The Infinity Puzzle.* (New York: Basic Books, 201), p. 348.

45. Lisa Randall. *Warped Passages.* (New York, Ecco, 2005), p. 258.

46. Edward Witten. "Foreword." appearing in *Supersymmetry* by Gordon Kane, (Cambridge, Mass, Perseus Publishing, 2000), p. xiii.

47. John Ellis (telephone interview with author), January 5, 2015.

48. John Ellis. "The Beautiful Physics of LHC Run 2." *Proceedings of Science.* December 8, 2014.
http://arxiv.org/abs/1412.2666

49. Y.A. Golfand and E.P. Likhtman. "Extension of Algebra of Poincare Group Generators and Violation of P Invariance. *JETP Letters.* 13 (1971), pp. 323-326.

    P. Ramond, "Dual theory for free fermions." *Physical Review D*, 3 (1971), pp. 2415-2418.

50. Corey S. Powell. "After the Higgs Boson: A Preview of Tomorrow's Radical Physics." *Discover* (online). October 11, 2013.
http://blogs.discovermagazine.com/outthere/2013/10/11/what-comes-after-the-higgs

51. Lisa Randall. *Warped Passages.* (New York, Ecco, 2005), p. 275.

52. Matt Strassler. "What is an 'Extra' Dimension? Some Examples." *Of Particular Significance (blog).* January 10, 2012. http://profmattstrassler.com

53. Matt Strassler (interview with the author at Harvard University). December 16, 2014.

54. Nima Arkani-Hamed (interview with the author in Cambridge, Mass.). April 1, 2015.

55. Nima Arkani-Hamed (interview with the author in Cambridge, Mass.). December 2, 2014.

56. Yifang Wang (interview with the author at Harvard University), December 2, 2014.

57. Matthew Schwartz, Harvard University (interview with the author at Harvard University), December 19, 2014.

58. Matthew Chalmers. "Q&A Martinus Veltman: Coming to terms with the Higgs." *Nature.* 490 (October 11, 2012), p. 511.

59. Nima Arkani-Hamed. After the Higgs Discovery: Where is Fundamental Physics Going?" Symposium, Tsinghua University, February 23, 2014.

60. Nathan Seiberg, Institute of Advanced Studies (telephone interview with author), March 31, 2015.

61. Nima Arkani-Hamed (interview with the author in Cambridge, Mass.). April 1, 2015.

Notes to Chapter 4

1. Chao-Lin Kuo, Stanford University (telephone interview with the author). April 18, 2014.

2. Meng Su, MIT (interview with the author), July 2, 2015.

3. Ibid.

4. Eugenie Samuel Reich. "Dark-matter hunt gets deep." *Nature.* 494 (February 21, 2013), pp. 291-2.

5. Adrian Cho. "Chinese team is catching up in hunt for dark matter." *Science* (online). August 26, 2014. http://news.sciencemag.org/physics/2014/08/chinese-team-catching-hunt-dark-matter

6. X. B. Cao, et al. "PandaX: A Liquid Xenon Dark Matter Experiment at CJPL." May 12, 2014. http://arxiv.org/abs/1405.2882

7. Ibid.

8. Ibid.

9. Eliza Strickland. "Deepest Underground Dark-Matter Detector to Start Up in China." *IEEE Spectrum* (online). January 29, 2014. http://spectrum.ieee.org/aerospace/astrophysics/ deepest-underground-darkmatter-detector-to-start-up-in-china

10. Eugenie Samuel Reich. "Dark-matter hunt gets deep." *Nature.* 494 (February 21, 2013), pp. 291-2.

11. X. B. Cao, et al. "PandaX: A Liquid Xenon Dark Matter Experiment at CJPL." http://arxiv.org/abs/1405.2882, May 12, 2014.

12. Charles p. Pierce. "Samuel Ting's space odyssey." *Boston Globe Magazine.* April 10, 2011.

13. Eric Hand. "Particle physics: Sam Ting's last fling." *Nature.* 455 (October 15, 2008), pp. 854-857.

14. Andrew Grant. "Sam Ting tries to expose dark matter's mysteries." *Science News.* 187 (March 21, 2015).

15. Kathryn Jepsen. "Pursuit of dark matter progresses at AMS." *Symmetry* (online). September 18, 2014.
    http://www.symmetrymagazine.org/article/september-2014/
    pursuit-of-dark-matter-progresses-at-ams

16. "Latest measurements from the AMS experiment unveil new territories in the flux of cosmic rays." *CERN Press Office* (online). September 2014.
    http://press.web.cern.ch/press-releases/2014/09/

17. John Matson. "Dark Matter Signal Possibly Registered on International Space Station. *Scientific American* (online). April 3, 2013.
    http://www.scientificamerican.com/article/dark-matter-ams

18. Ibid.

19. "Latest measurements from the AMS experiment unveil new territories in the flux of cosmic rays." *CERN Press Office* (online).
    http://press.web.cern.ch/press-releases/2014/09/

20. Lisa Grossman. "Spaceborne dark matter hunter sees telltale antimatter." *New Scientist* (online). April 3, 2013.
    http://www.newscientist.com/article/
    dn23342-spaceborne-dark-matter-hunter-sees-telltale-antimatter.html

21. Sarah Charley. "AMS results create cosmic ray puzzle." *Symmetry* (online). April 15, 2015.
    http://www.symmetrymagazine.org/article/april-2015/
    ams-results-create-cosmic-ray-puzzle

22. Dennis Overbye. "A Costly Quest for the Dark Heart of the Cosmos." *New York Times* (online). November 16, 2010.
    http://www.nytimes.com/2010/11/17/science/space/17dark.html

23. Clara Moskowitz. "Antimatter Hunter: Q&A with Nobel Laureate Sam Ting." *Space.com.* May 15, 2011.
    http://www.space.com/11671-shuttle-alpha-magnetic-spectrometer-ting.html

24. David A. Weintraub. *How Old is the Universe?* (Princeton, New Jersey: Princeton University Press, 2011), p. 285.

25. John Bahcall. "Solving the Mystery of the Missing Neutrinos." *Nobelprize.org.*
    http://www.nobelprize.org/nobel_prizes/themes/physics/bahcall

26. Steven Weinberg. "The Revolution That Didn't Happen." *New York Review of Books.* 45 (October 8, 1998), pp. 48-52.

27. John Bahcall. "Solving the Mystery of the Missing Neutrinos."

28. Ming-Chung Chu, Chinese University of Hong Kong (interview with the author at the Hong Kong University of Science and Technology), August 14, 2014.

29. Matt Strassler. "Neutrino Types and Neutrino Oscillations." *Of Particular Significance* (blog). October 5, 2011, http://profmattstrassler.com

30. Jon Butterworth. *Most Wanted Particle.* (New York: The Experiment, 2014), p. 213.

31. "Daya Bay collaboration observes a new kind of neutrino oscillation." *CERN Courier.* April 27, 2012.
http://cerncourier.com/cws/article/cern/49336

32. Yifang Wang. "Daya Bay Neutrino Experiment and the Future." (lecture at Harvard University). December 2, 2014.

33. Adrian Cho. "Physicists in China Nail a Key Neutrino Measurement." *Science* (online). March 8, 2012.
http://news.sciencemag.org/asia/2012/03/
physicists-china-nail-key-neutrino-measurement

34. Eugenie Samuel Reich. "Neutrino oscillations measure with record precision." *Nature.com* newsblog. March 8, 2012.
http://blogs.nature.com/news/2012/03/
neutrino-oscillations-measured-with-record-precision.html

35. Kathryn Jepsen. "Daya Bay furthers neutrino knowledge." *Symmetry* (online). August 22, 2013.
http://www.symmetrymagazine.org/article/august-2013/
daya-bay-furthers-neutrino-knowledge

36. Ming-Chung Chu (interview with the author at the Hong Kong University of Science and Technology), August 14, 2014.

37. Jiao Li. "China to build a huge underground neutrino experiment." *Physics World* (online). March 4, 2014.
http://physicsworld.com/cws/article/news/2014/mar/24/
china-to-build-a-huge-underground-neutrino-experiment

38. Yifang Wang. "Daya Bay Neutrino Experiment and the Future" (lecture at Harvard University). December 2, 2014.

39. IHEP. "Groundbreaking at JUNO" (press release). January 10, 2015.

40. Yifang Wang, IHEP (interview with the author at Harvard University), December 2, 2014.

41. Jiao Li. "China to build a huge underground neutrino experiment."

42. Kendra Snyder. "Catching Neutrinos in China." *Symmetry*. 3 (October-November 2006), pp. 16-21.

43. Adrian Cho. "Physicists in China Nail a Key Neutrino Measurement." *Science* (online). March 8, 2012. http://news.sciencemag.org/asia/2012/03/physicists-china-nail-key-neutrino-measurement

44. Eugenie Samuel Reich. "Neutrino oscillations measure with record precision." *Nature* (online). March 8, 2012. http://blogs.nature.com/news/2012/03/neutrino-oscillations-measured-with-record-precision.html

45. Adrian Cho. "Physicists in China Nail a Key Neutrino Measurement."

46. Adrian Cho. "Key Neutrino Measurement Signal's China's Rise." *Science*. 365 (March 16, 2012), pp. 1287-88.

47. Charles Petit. "Heart of the matter." *Science News* (online). 183.2 (January 10, 2013). https://www.sciencenews.org/article/heart-matter

48. "2014 W.K.H. Panofsky Prize in Experimental Particle Physics Recipient." *APS Physics*. http://www.aps.org/programs/honors/prizes

49. Ming-Chung Chu (interview). August 14, 2014.

50. Gerard 't Hooft. Utrecht University (telephone interview with the author), April 24, 2015.

51. Cong Cao. "Chinese Science and the 'Nobel Prize Complex.' " *Minerva*. 42 (2004), pp. 151-172.

52. Mark Walker (editor). *Science and Ideology: A Comparative History*. Routledge (Abingdon, Oxford, UK: Routledge, 2003), p. 56.

53. Stuart Schram. *The Thought of Mao Tse-Tung*. Cambridge University Press (Cambridge, UK: 1989), p. 168.

54. Dennis Overbye. "China Pursues Major Role in Particle Physics." *New York Times* (online). December 5, 2006. http://www.nytimes.com/2006/12/05/science/05china.html

55. Zuoyue Wang. "U.S.-China scientific exchange." *Historical Studies in the Physical and Biological Sciences*. 30 (1999), pp. 249-277.

56. "China's IHEP celebrates its first 40 years." *CERN Courier*. November 20, 2013, http://cerncourier.com/cws/article/cern/55335/3

57. Zuoyue Wang. "U.S.-China scientific exchange." *Historical Studies in the Physical and Biological Sciences*. 30 (1999), pp. 249-277.

58. Wenxian Zhang. "Ting, Samuel C.C." *Asian American History and Culture: An Encyclopedia* (edited by Huping Ling and Alan W. Austin). Routledge (Abingdon, Oxon, UK: 2015), p. 237.

59. Dennis Overbye. "China Pursues Major Role in Particle Physics." *New York Times* (online). December 5, 2006.
http://www.nytimes.com/2006/12/05/science/05china.html

60. Wolfgang K.H. Panofsky. *Panofsky on Physics, Politics, and Peace.* Springer (New York, 2007), pp. 129-134.

61. Zuoyue Wang. "U.S.-China scientific exchange." *Historical Studies in the Physical and Biological Sciences.* 30 (1999), pp. 249-277.

62. Wolfgang K.H. Panofsky. *Panofsky on Physics, Politics, and Peace.* Springer (New York, 2007), pp. 129-134.

63. Liu Huaizu (ed.) *The Beijing Electron Positron Collider.* (Beijing, IHEP, 1994), p. 40

64. Dennis Overbye. "China Pursues Major Role in Particle Physics." *New York Times* (online). December 5, 2006.
http://www.nytimes.com/2006/12/05/science/05china.html

65. Chuang Zhang, IHEP (interview with the author at IHEP). August 8, 2014.

66. Ibid.

67. Liu Huaizu (ed.) *The Beijing Electron Positron Collider.* (Beijing, IHEP, 1994), pp. 93, 231

68. Z. G. Zhao. "BES Recent Results and Future Plans." *ArXiv.org.* December 22, 2000.
http://arxiv.org/abs/hep-ex/0012056

69. Dennis Overbye. "China Pursues Major Role in Particle Physics." *New York Times* (online). December 5, 2006.
http://www.nytimes.com/2006/12/05/science/05china.html

70. Martin L. Perl. "The Discovery of the Tau Lepton." SLAC-PUB-5937, September 1992.

71. Fred Harris, University of Hawaii (telephone interview with the author). February 27, 2015.

72. Dennis Overbye. "China Pursues Major Role in Particle Physics." *New York Times* (online). December 5, 2006.
http://www.nytimes.com/2006/12/05/science/05china.html

73. Wolfgang K.H. Panofsky. *Panofsky on Physics, Politics, and Peace.* Springer (New York, 2007), pp. 133-134.

74. "BEPC II celebrates the first collision events." *CERN Courier* (online). August 18, 2008. http://cerncourier.com/cws/article/cern/35434

75. Kelen Tuttle. "Chasing charm in China." *Symmetry.* May 2009, pp. 15-19.

76. Ibid.

77. "BESIII observes new mystery particle." *CERN Courier* (online). April 26, 2013. http://cerncourier.com/cws/article/cern/53072

78. Hesheng Chen. "China agrees upgrade of its particle collider. *CERN Courier* (online). March 31, 2003. http://cerncourier.com/cws/article/cern/28824

79. Xiaoyan Shen (interview with the author at IHEP). August 5, 2014.

80. Eric Swanson. "Viewpoint: New Particle Hints at Four-quark Matter." *APS Physics* (online). June 17, 2013. http://physics.aps.org/articles/v6/69

81. Devin Powell. "Quark quartet opens fresh vista on matter." *Nature.* 498 (June 20, 2013), pp. 280-281.

82. Lisa Grossman. "What a new jumbo particle reveals about extreme matter." *New Scientist* (online). June 24, 2013. http://www.newscientist.com/article/ dn23726-what-a-new-jumbo-particle-reveals-about-extreme-matter.html

83. "High Energy Physicists Predict New Family of Four-Quark Objects." *Kaunānā. The Research Publication of the University of Hawaii at Manoa.* November 6, 2013.

84. Kelen Tuttle. "Chinese collider expands particle zoo." *Symmetry* (online). December 9, 2013. http://www.symmetrymagazine.org/article/chinese-collider-expands-particle-zoo

85. Xiaoyan Shen (interview with the author at IHEP). August 5, 2014.

86. Fred Harris, University of Hawaii (telephone interview with the author). February 27, 2015.

87. Yifang Wang, IHEP (e-mail correspondence with the author), April 4, 2015.

88. Robbert Dijkgraaf, Institute for Advanced Study (telephone interview with the author). April 20, 2015.

89. Chuang Zhang (interview with the author at IHEP). August 8, 2014.

90. Fred Harris, University of Hawaii (telephone interview with the author). February 27, 2015.

Notes to Chapter 5

1. Dennis Overbye. "Asking a Judge to Save the World and Maybe a Whole Lot More." *New York Times* (online). March 29, 2008.
http://www.nytimes.com/2008/03/29/science/29collider.html

   "First beam in the LHC." *CERN Press Office* (online). September 10, 2008.
   http://press.web.cern.ch/press-releases/2008/09/first-beam-lhc-accelerating-science

2. CERN. "Synchrotron Radiation." *Taking a Closer Look at LHC.*
http://www.lhc-closer.es

3. Alain Blondel, et al, "Accelerators for a Higgs Factory: Linear vs. Circular" (HF2012). *Report of the ICFA Beam Dynamics Workshop.* February 15, 2013, arXiv: 1302.3318.

4. Yifang Wang (e-mail communication with the author), May 13, 2015.

5. "Future Circular Collider Study." *CERN* (online). January 23, 2014.
https://fcc.web.cern.ch

6. Xinchou Lou, IHEP (telephone interview with the author), August 25, 2014.

7. Fabiola Gianotti. The Second Annual Large Hadron Collider Physics Conference. Columbia University. June 6, 2014.

8. Adrian Cho. "Mega-Doughnuts." *Science* (online). February 6, 2014.
http://news.sciencemag.org/europe/2014/02/
mega-doughnuts-cern-study-plan-100-kilometer-atom-smashers

9. Jie Gao, IHEP (interview with the author). August 6, 2014.

10. Michael Chanowitz. "What if there is no Higgs boson?" *ATLAS News* (online). November 30, 2011.
http://www.atlas.ch/news/2011/what-if-there-is-no-higgs-boson.html

11. Sean Carroll. "Why We Need the Higgs, or Something Like It." *Discover* (online). June 14, 2011.
http://blogs.discovermagazine.com/cosmicvariance/2011/06/14/
why-we-need-the-higgs-or-something-like-it

12. Dennis Overbye. "An Ambassador for Physics Is Shifting His Mission." *New York Times* (online). January 12, 2015.
http://www.nytimes.com/2015/01/13/science/
departing-leader-of-cern-ponders-uncertainties-that-lie-ahead.html

13. Michael Peskin. "I want the ILC!" *InternationalLinearColliderTV.* November 14, 2014.

14. The CEPC-SPPC Study Group. *Preliminary Conceptual Design Report: Physics and Detector.* March 2015, p. 9.

15. The CEPC-SPPC Study Group. *Preliminary Conceptual Design Report: Volume II — Accelerator.* March 2015, p. 31.

16. Ashutosh Kotwal, Duke University/Fermilab (telephone interview with the author), April 13, 2015.

17. The CEPC-SPPC Study Group. *Preliminary Conceptual Design Report: Physics and Detector.* March 2015, p. 39.

18. Ashutosh Kotwal, Duke University/Fermilab (telephone interview with the author), April 13, 2015.

19. The CEPC-SPPC Study Group. *Preliminary Conceptual Design Report: Physics and Detector.* March 2015, p. 39.

20. Nathan Seiberg, Institute for Advanced Study (telephone interview with the author), March 31, 2015.

21. Michael Peskin, SLAC (e-mail communication with the author), May 25, 2015.

22. A. D. Sakharov. "Violation of CP Invariance, C Asymmetry, and Baryon Asymmetry of the Universe." *ZhETF Pis'ma.* 5 (1967), pp. 32-35.
    http://www.jetpletters.ac.ru/ps/1643/article_25089.pdf

23. "BaBar finds direct CP violation in B decays." *CERN Courier* (online). September 5, 2004.
    http://cerncourier.com/cws/article/cern/29124

24. Henry Tye, Hong Kong University of Science and Technology (e-mail communication with the author), May 16, 2015.

25. Ashutosh Kotwal, Duke University/Fermilab (telephone interview with the author), April 13, 2015.

26. Michael Peskin, SLAC (e-mail communication with the author), May 25, 2015.

27. Ashutosh Kotwal, Duke University/Fermilab (telephone interview with the author), April 13, 2015.

28. Xinchou Lou, IHEP (telephone interview with the author), August 25, 2014.

29. David Gross, University of California, Santa Barbara (telephone interview with the author), April 16, 2015.

30. "Status of the project." *International Linear Collider.*
    https://www.linearcollider.org/ILC/What-is-the-ILC/Status-of-the-project

31. Michael Peskin, SLAC (telephone interview with the author), March 31, 2015.

32. Giorgio Ambrosio, et al. "Design Study for a Staged Very Large Hadron Collider." *Fermilab-TM-2149*. June 4, 2001.
http://www.slac.stanford.edu/cgi-wrap/getdoc/slac-r-591.pdf

33. James Glanz. "Physicists Unite, Sort of, on Next Collider." *New York Times* (online). July 10, 2001.
http://www.nytimes.com/2001/07/10/
science/physicists-unite-sort-of-on-next-collider.html

34. Eugenie Samuel Reich. "Physicists plan to build a bigger LHC." *Nature*. 503 (November 14, 2014), p. 177.

35. Roland Pease. "CERN considers building huge physics machine." *BBC News* (online). February 18, 2014.
http://www.bbc.com/news/science-environment-26250716

36. Yifang Wang, IHEP (telephone interview with the author), April 3, 2015.

37. Nathan Seiberg, IAS (telephone interview with the author), March 31, 2015.

38. Nima Arkani-Hamed, IAS (interview with author in Cambridge, Mass.), April 1, 2015.

39. Luciano Maiani, University of Rome (telephone interview with the author). April 22, 2015.

40. "U.S. Joins the World in a New Era of Research at the Large Hadron Collider." *Brookhaven National Laboratory Newsroom* (online). June 3, 2015.
http://www.bnl.gov/newsroom/news.php?a=11734

41. Michael Peskin, SLAC (telephone interview with the author), March 31, 2015.

42. Gerard 't Hooft. Utrecht University (telephone interview with the author), April 24, 2015.

43. Yifang Wang, IHEP (telephone interview with the author), April 3, 2015.

44. Henry Tye (interview with the author at the Hong Kong University of Science and Technology), August 12, 2014.

45. Michael Peskin, SLAC (interview with author at IHEP), August 8, 2014

46. The CEPC-SPPC Study Group. *Preliminary Conceptual Design Report: Volume II -- Accelerator*. March 2015, p. 325.

47. Nima Arkani-Hamed, IAS (interview with author in Cambridge, Mass.), April 1, 2015.

48. Sarah Charley. Test magnet reaches 13.5 tesla—a new CERN record." *CERN* (online). November 18, 2013.
http://home.web.cern.ch/about/updates/2013/11/
test-magnet-reaches-135-tesla-new-cern-record

49. Yifang Wang. "Daya Bay Neutrino Experiment and the Future" (lecture at Harvard University). December 2, 2014.

50. Yifang Wang, (interview with the author at IHEP), August 4, 2014.

51. S. Myers and W. Schnell. "Preliminary Performance Estimates for a LEP Proton Collider. *LEP Note 440*. November 4, 1983.

52. Jon Butterworth. "The future of particle physics?" *The Guardian* (online). February 22, 2014. http://www.theguardian.com/science/life-and-physics/ 2014/feb/22/the-future-of-particle-physics

53. Qing Qin, IHEP (interview with the author), August 7, 2014.

54. Yifang Wang, IHEP (telephone interview with the author), April 3, 2015.

55. Qing Qin, IHEP (interview with the author), August 7, 2014.

56. Xinchou Lou. "The CEPC-SppC Study Group in China: Introduction, Status and Future Plans." February 24, 2014.

57. Yifang Wang, IHEP (telephone interview with the author), April 3, 2015.

58. Nima Arkani Hamed, IAS (interview with the author at IHEP), August 8, 2014.

59. Xinchou Lou, IHEP (telephone interview with the author), August 25, 2014.

60. Nima Arkani-Hamed, IAS (interview with author in Cambridge, Mass.), April 1, 2015.

61. Nima Arkani-Hamed. The Second Annual Large Hadron Collider Physics Conference. Columbia University. June 6, 2014.

62. Robbert Dijkgraaf, IAS (telephone interview with the author). April 20, 2015.

63. Yifang Wang, IHEP (telephone interview with the author), April 3, 2015.

64. David Gross, University of California, Santa Barbara (telephone interview with the author), April 16, 2015.

65. Luciano Maiani, University of Rome (telephone interview with the author). April 22, 2015.

66. Robbert Dijkgraaf, IAS (telephone interview with the author). April 20, 2015.

67. David Gross, University of California, Santa Barbara (telephone interview with the author), April 16, 2015.

68. Nima Arkani-Hamed, IAS (interview with author in Cambridge, Mass.), April 1, 2015.

69. Yifang Wang, IHEP (e-mail communication with the author), May 26, 2015.

70. Yifang Wang. "After the Higgs Discovery: Where is Fundamental Physics Going?" Symposium, Tsinghua University, February 23, 2014.

71. Yifang Wang, IHEP (telephone interview with the author), April 3, 2015.

Notes to Chapter 6

1. Leon Lederman and Dick Teresi. *The God Particle.* (Boston and New York: Houghton Mifflin Harcourt, 2006), p. 124.

2. "This Month in Physics History — Discovery of Nuclear Fission." *APS Physics* (online). 16 (December 2007. http://www.aps.org/publications/apsnews/200712/physicshistory.cfm

3. David Gross. "After the Higgs Discovery: Where is Fundamental Physics Going?" (Symposium). Tsinghua University, Beijing, February 23, 2014.

4. Steven Weinberg. *Dreams of a Final Theory.* (New York, Vintage Books, 1992), p. 282.

5. Kate Allen. "Basic science is the centre of gravity." *Toronto Star* (online). November 25, 2014. http://www.thestar.com/news/world/2014/11/24/ basic_science_is_the_centre_of_gravity_says_particle_physics_lab_chief.html

6. Philip Burrows. "Particle Accelerators 1." The 2d Institute of Advanced Studies School on Particle Physics and Cosmology and Implications for Technology. Nanyang Technical University, February 2, 2015.

7. Rolf Heuer. "One Day, Sir, You May Tax It." *CERN Bulletin* (online). February 14, 2011. https://cds.cern.ch/journal/CERNBulletin/2011/07

8. Dennis Overbye. "Collider Sets Record, and Europe Takes U.S.'s Lead. *New York Times* (online). December 9, 2009. http://www.nytimes.com/2009/12/10/science/10collide.html

9. Ashutosh Kotwal, Duke University/Fermilab (telephone interview with the author), April 13, 2015.

10. Tim Berners-Lee. "Information Management: A Proposal." March 1989. http://www.w3.org/History/1989/proposal.html

11. "Time Berners-Lee's proposal." *CERN* (online). http://info.cern.ch/Proposal.html

12. Michael Moore. "More than Three Billion People Worldwide Now Have Broadband." *MSN TechWeekEurope.* January 26, 2015. http://a.msn.com/01/en-gb/AA8BcD1?ocid=se

13. "History of the Web." *World Wide Web Foundation.*
    http://webfoundation.org/about/vision/history-of-the-web

14. Sean Carroll. *The Particle at the End of the Universe.* (New York: Dutton, 2012), p. 274.

15. "Large Hadron Collider restarts after two years." *University of Cambridge Research News* (online). April 7, 2015.
    http://www.cam.ac.uk/research/news/large-hadron-collider-restarts-after-two-years

16. "Computing." *CERN Accelerating Science* (online).
    http://home.web.cern.ch/about/computing

17. "The Worldwide LHC Computing Grid" *CERN Accelerating Science* (online).
    http://home.web.cern.ch/about/computing/worldwide-lhc-computing-grid

18. Joseph Incandela. "After the Higgs Discovery: Where is Fundamental Physics Going?" (Symposium). Tsinghua University, Beijing, February 23, 2014.

19. Luciano Maiani, University of Rome (telephone interview with the author), April 22, 2015.

20. Luciano Maiani. "After the Higgs Discovery: Where is Fundamental Physics Going?" (Symposium). Tsinghua University, Beijing, February 23, 2014.

21. Hitoshi Murayama. "After the Higgs Discovery: Where is Fundamental Physics Going?" (Symposium). Tsinghua University, Beijing, February 23, 2014.

22. Henry Tye, Hong Kong University of Science and Technology (e-mail communication with the author), April 5, 2015.

23. Henry Tye, Hong Kong University of Science and Technology (e-mail communication with the author), June 2, 2015.

24. Nima Arkani-Hamed, IAS (interview with author in Cambridge, Mass.), April 1, 2015.

25. Yifang Wang, IHEP (interview with the author), August 4, 2014.

26. Xinchou Lou. "The CEPC-SppC Study Group in China: Introduction, Status and Future Plans." February 24, 2014.

27. Weiren Chou, Fermilab (interview with the author at IHEP), August 4, 2014.

28. Ashutosh Kotwal Ashutosh Kotwal, Duke University/Fermilab (telephone interview with the author), April 13, 2015.

29. Lisa Randall. *Higgs Discovery.* (New York, Ecco, 2013), p. 39.

30. Joseph Incandela. "After the Higgs Discovery: Where is Fundamental Physics Going?" (Symposium). Tsinghua University, Beijing, February 23, 2014.

31. Robbert Dijkgraaf, IAS (telephone interview with the author). April 20, 2015.

32. Toni Feder. "Particle Fever: Filming the hunt for the Higgs boson." *Physics Today* (online). January 30, 2014.
    http://scitation.aip.org/content/aip/magazine/physicstoday/news/10.1063/PT.5.9007

33. Clara Moskowitz. "Antimatter Hunter." *Space.com.* May 15, 2011.
    http://www.space.com/11671-shuttle-alpha-magnetic-spectrometer-ting.html

34. Robbert Dijkgraaf, IAS (telephone interview with the author). April 20, 2015.

35. Frank Wilczek. " 'National greatness' versus real national greatness." *Science News.* October 11, 2008, p. 32.

Index